Table of Contents

Preface

The primary goal of this book is to help the reader understand automotive electronic systems. Many readers who formerly performed routine maintenance and tune-up on their own cars are confused by the seemingly complex and unfamiliar electronic devices which they find under the hood and behind the instrument panel. This book is aimed at those readers who have developed a curiosity about these devices and want to understand their functions and how they are used in the automobile.

The understanding of the function of automotive electronic systems progresses step by step. After an overview of automotive fundamentals and how systems are analyzed using a systems approach, two chapters reinforce the reader's knowledge of semiconductor devices, integrated circuits, digital circuits and how microcomputers are used in instrumentation control. With this background, how physical quantities are converted to electrical signals by sensors, how electronic systems perform automotive functions by processing the signals, and how electrical signals are converted back to physical action by actuators completes the discussion through the next six chapters. Engine control, motion control, instrumentation and the future are major topics.

The technology being developed in the U.S. automobile industry is relatively sophisticated and the explanations presented in the technical engineering publications often are not easy to understand. We have attempted to sort out the important, basic issues in automotive electronics and to present them in an easy to read and easy to understand language. Because of this, some explanations may seem overly simplified to some readers. For those readers who would like more details about these complex subjects, the best source of material is the literature which is available from the Society of Automotive Engineers (SAE).

This book, like the others in the series, is designed to build understanding step-by-step. Try to master each chapter before going on to the next one. Some readers experienced in electronics may feel they can bypass Chapters 3 and 4. If you are one of these, that's fine; however, they do serve as an excellent review. A quiz is provided at the end of each chapter for personal evaluation of progress. Answers are included.

A glossary and index are provided to aid in using and understanding the material and finding the subjects of interest.

Automotive electronic technology is changing rapidly; therefore, future automotive electronic systems will probably differ from the systems discussed in this book. However, with the understanding gained by reading this book, the reader should be able to follow developments to be made in the future in this exciting field.

W. B.
N. M.

Automotive Fundamentals

Allan, Carolyn and the kids are taking a cross-country trip. The time is in the not-too-distant future. No one is driving the car because Allan has turned on the auto pilot and the car is being driven automatically by following a signal from a wire buried just beneath the highway surface. The signal controls the steering and speed of the car.

Allan and Carolyn's car is in a string of vehicles all cruising at an absolutely controlled speed of 90 km/hr (about 55 mph). The spacing between the vehicles is a safe 50 m (about half the length of a football field). This spacing is also controlled automatically. Occasionally, a vehicle enters or leaves the string of vehicles from one of the automatic entry-exit lanes.

Allan and family are playing a game and listening to music over the computer-controlled entertainment system. Suddenly the music is interrupted and a message comes over the loudspeakers: "You have 75 km to empty tank. Your programmed destination is beyond this range. Recommend refueling at the next fuel stop which comes in 37 km. Also, check the right front tire as the air pressure is low." Then the music returns.

This message is delivered in a natural sounding synthesized voice by the computer controlled vehicle instrumentation and monitoring system. In addition, this system continuously displays vehicle speed, fuel quantity range to empty, position along the programmed route, and time and date on a visual display similar to a small television screen.

The use of micro-electronics will play a major role in automotive technology in the not-too-distant future.

Does this sound far fetched? Impossible? Well, it is neither. Most of the events described are technically possible and have been tried experimentally. They have been made possible by the rapid technical developments in solid-state electronics.

USE OF ELECTRONICS IN THE AUTOMOBILE

Electronics has been relatively slow in coming to the automobile primarily because of the relationship between the added cost and the benefits. Historically, the first electronics was introduced into the commercial automobile during the decade 1930–1940 in the form of automobile radio receivers. There were a few attempts to introduce electronic ignition and electronically controlled fuel injection during the late 1950's and early 1960's. However, customers did not particularly want these options so they were discontinued from production automobiles.

Environmental regulations and an increased need for economy have resulted in electronics being used within a number of automotive systems.

Two major events occurred during the 1970's which started the trend toward the use of modern electronics in the automobile: 1) the introduction of government regulations for exhaust emissions and fuel economy which required better control of the engine than was possible with the methods being used; and 2) the development of relatively low cost per function solid-state digital electronics which could be used for engine control.

Electronics are being used now in the automobile and probably will be used even more in the future. Some of the present and potential applications for electronics are:

1) electronic engine control for minimizing exhaust emissions and maximizing fuel economy
2) instrumentation for measuring vehicle performance parameters and for diagnosis of on-board system malfunctions
3) driveline control
4) vehicle motion control
5) safety and convenience

Many of these applications of electronics will be discussed in this book.

ABOUT THIS CHAPTER

This chapter is intended to give the reader a general overview of the automobile with emphasis on the basic operation of the engine. This will provide the reader with the background to see how electronic controls have been and will be applied. The discussion is simplified to provide the reader with just enough information to understand automotive electronics. Readers who want to know the mechanics of an automobile in more detail are referred to the many books written for that purpose.

THE AUTOMOBILE

The important systems of the automobile are illustrated in *Figure 1-1* and include:

1) engine
2) drivetrain
3) instrumentation

4) suspension
5) steering
6) brakes

In this figure, the frame or chassis upon which the body is mounted is supported by the wheel suspension system. (Some passenger cars are constructed so that the auto body and chassis are not separate). Moreover, many of the newer cars are being designed with front wheel drive. Nevertheless, this figure provides a convenient reference for discussing automotive fundamentals.

We will see in this book that whenever electronics is used, significant improvements have been achieved in automobile performance.

**Figure 1-1.
Systems of the
Automobile**

*Engine
mounted
transversely
for front
wheel drive.*

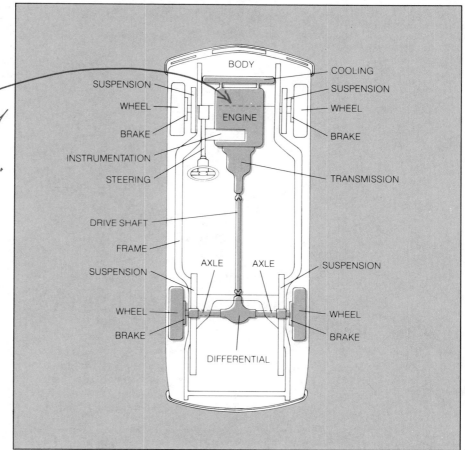

THE ENGINE

The engine in an automobile provides all the power for moving the automobile, for the hydraulic and pneumatic systems, and for the electrical system. A variety of engine types have been produced, but one class of engine is used most; i.e., the internal combustion, piston type, four stroke/cycle, gasoline fueled, spark ignited, liquid-cooled engine. This engine will be referred to in this book as the spark ignited or SI engine. A typical SI engine is depicted in *Figure 1-2*.

The major components of the engine include:

1) engine block	7) cylinder head
2) cylinder	8) valves
3) crankshaft	9) intake system
4) pistons	10) ignition system
5) connecting rods	11) exhaust system
6) camshaft	12) cooling system

Figure 1-2.
Cutaway View of a Six-
Cylinder, Overhead-
Valve, In-Line Engine
(Source: Crouse)

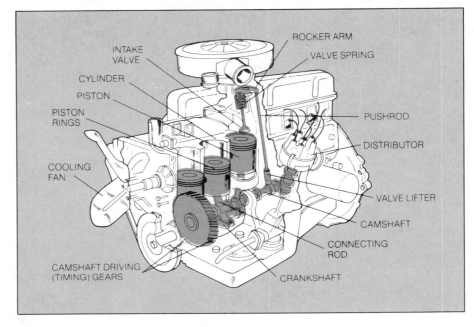

Engine Block

Conventional internal combustion engines convert
the movement of pistons to
the rotational energy used
to drive the wheels.

    The cylinders are cast in the engine block and machined to a smooth
finish. The pistons have rings which provide a tight sliding seal against the
cylinder wall. The pistons are connected to the crankshaft by connecting rods
as shown in *Figure 1-3*. The crankshaft converts the up and down motion of the
pistons to the rotary motion needed to drive the wheels.

Figure 1-3.
Piston Connection to
Crankshaft
(Source: Crouse)

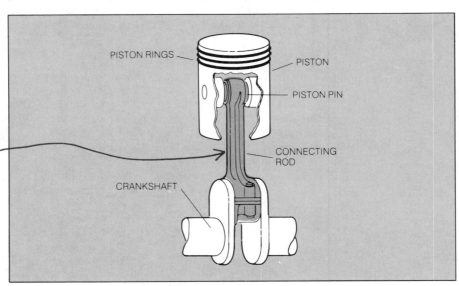

The rod that provides the connection between crankshaft and piston.

Cylinder Head

The cylinder head contains an intake and exhaust valve for each cylinder. When both valves are closed, the head seals the top of the cylinder while the piston rings seal the bottom of the cylinder.

The valves are operated by off center (eccentric) cams on the camshaft which is driven by the crankshaft as shown in *Figure 1-4*. The camshaft rotates at exactly half the crankshaft speed because a complete cycle of any cylinder involves two complete crankshaft rotations and only one sequence of opening and closing of the associated intake and exhaust valves. The valves are normally held closed by powerful springs. When the time comes for a valve to open, the lobe on the cam forces the pushrod upward against one end of the rocker arm. The other end of the rocker arm goes downward and forces the valve open. (Note: Some engines have the camshaft above the head so the pushrods are eliminated. This is called an overhead cam engine.)

Figure 1-4.
Valve Operating
Mechanism
(Source: Crouse)

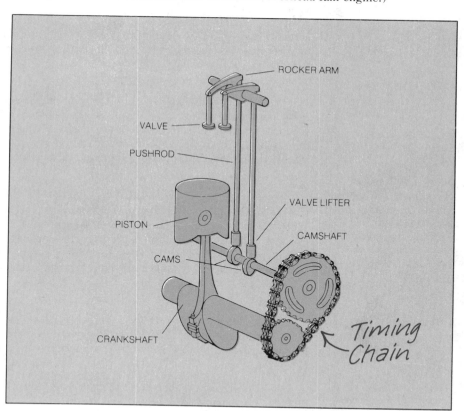

The Four-Stroke Cycle

Conventional SI engines operate using four "strokes", either an up or down movement of each piston. These strokes are named intake, compression, power, and exhaust.

The operation of the engine can be understood by considering the actions in any one cylinder during a complete cycle of the engine. One complete cycle in the 4 stroke/cycle SI engine requires two complete rotations of the crankshaft. As the crankshaft rotates, the piston moves up and down in the cylinder. In the two complete revolutions of the crankshaft that make up one cycle, there are 4 separate strokes of the piston from the top of the cylinder to the bottom or from the bottom to the top. *Figure 1-5* illustrates the 4 strokes for a 4 stroke/cycle SI engine which are called:

1) intake
2) compression
3) power
4) exhaust

There are two valves for each cylinder. The left valve in the figure is called the intake valve and the right valve is called the exhaust valve. The intake valve is normally larger than the exhaust valve. Note that the crankshaft is assumed to be rotating in a clockwise direction. The action of the engine during the 4 strokes is described below.

Intake

During the intake stroke (*Figure 1-5a*), the piston is moving from top to bottom and the intake valve is open. As the piston moves down, a vacuum or suction is created which draws a mixture of air and vaporized gasoline through the intake valve into the cylinder. The intake valve is closed after the piston reaches the bottom. This position is normally called bottom dead center (BDC).

Compression

The intake stroke draws a combustible mixture of air and gasoline into the cylinder; the compression stroke compresses this mixture in preparation for combustion.

During the compression stroke (*Figure 1-5b*), the piston moves upward and compresses the fuel and air mixture against the cylinder head. When the piston is near the top of this stroke, the ignition system produces an electrical spark at the tip of the spark plug. [The top of the stroke is normally called top dead center (TDC).] The spark ignites the air-fuel mixture and the mixture burns rapidly causing a rapid and extreme rise in the pressure in the cylinder.

Power

During the power stroke (*Figure 1-5c*), the high pressure created by the burning mixture forces the piston downward. It is only during this stroke that actual usable power is generated by the engine.

Exhaust

The power stroke ignites the mixture and creates a downward force on the surface of the piston. The exhaust stroke forces the exit of burned gases from the cylinder, in preparation for the next intake stroke.

During the exhaust stroke (*Figure 1-5d*), the piston is again moving upward. The exhaust valve is open and the piston forces the burned gases from the cylinder through the exhaust port into the exhaust system and out the tailpipe into the atmosphere.

Figure 1-5.
The 4 Strokes of a 4-
Stroke/Cycle SI Engine

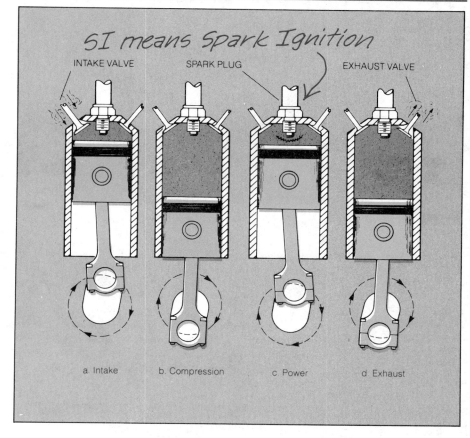

SI means Spark Ignition

INTAKE VALVE   SPARK PLUG   EXHAUST VALVE

a. Intake   b. Compression   c. Power   d. Exhaust

This four-stroke cycle is repeated continuously as the crankshaft rotates. In a single cylinder engine, power is produced only during the power stroke which is only one-quarter of the cycle. In order to maintain crankshaft rotation during the other three-quarters of the cycle, a flywheel is used. The flywheel is a relatively large, heavy, circular, object which is connected to the crankshaft. The primary purpose of the flywheel is to provide inertia to keep the crankshaft rotating during the three non-power-producing strokes of the piston.

Each piston on a 4-cycle SI engine produces actual power during just one out of four cycles.

In a multi-cylinder engine, the power strokes are staggered so power is produced during a larger fraction of the cycle than for a single cylinder engine. In a four cylinder engine, for example, power is produced almost continually by the separate power strokes of the 4 cylinders. The shaded regions of *Figure 1-6* indicate which cylinder is producing power for each 180 degrees of crankshaft rotation. (Remember that one complete engine cycle requires two complete crankshaft rotations of 360 degrees each for a total of 720 degrees.)

**Figure 1-6.
Power Pulses from a
4-Cylinder Engine**

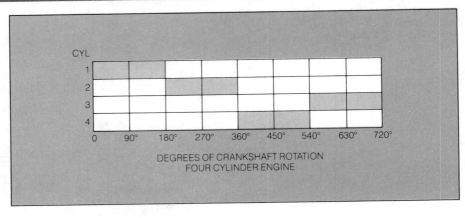

INTAKE SYSTEM

    The intake system consists of a carburetor and an assembly of passageways called the intake manifold. The carburetor mixes the air and fuel and the intake manifold routes the mixture to the cylinders as shown in *Figure 1-7*.

    The *proportion* of air and fuel in the mixture delivered to the cylinder is basically controlled by the size and shape of the bore of the carburetor and the size and shape of the metering rods and seats (sometimes called "jets") in the carburetor. The proportion of air and fuel in the mixture is expressed by the ratio of the mass (weight) of air to the mass (weight) of fuel. This ratio is appropriately called the air/fuel ratio. In normal operation, the air/fuel ratio varies in the range of 12:1 to 17:1.

A combination of carburetor and intake manifold are used to route precise amounts of fuel and air to individual cylinders.

**Figure 1-7.
Carburetor and Intake
Manifold**

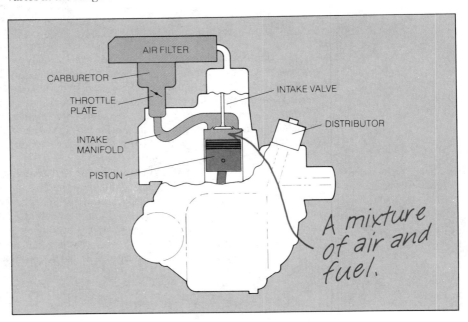

The *amount* of the air and fuel mixture delivered to the engine is controlled by the throttle plate. The throttle plate, which acts as an air flow control valve, is controlled by the accelerator pedal.

It is beyond the scope of this book to discuss the details of carburetor operation, but it is important to realize that each carburetor is designed to deliver the correct proportion of air and fuel for a specific engine.

IGNITION SYSTEM

The ignition system provides an electric spark during the compression stroke which ignites the air-fuel mixture. This spark consists of an electric arc across the electrodes of the spark plug. The operation of the ignition system can be explained by describing the operation of a traditional non-electronic system. (Engine performance has already been greatly improved using electronics in the ignition system. This will be discussed in a later chapter.)

Spark

A typical spark plug configuration is shown in *Figure 1-8*. The spark plug consists of a pair of electrodes, called the center and ground electrodes, separated by a gap. The gap size is important and is specified for each engine. The gap may be 0.025 inch (0.6mm) for one engine and 0.040 inch (1mm) for another engine. The center electrode is insulated from the ground electrode and the metallic shell assembly. The ground electrode is at electrical ground potential because one terminal of the battery is connected to the engine block and frame. This is called the ground connection.

**Figure 1-8.
Spark Plug
Configuration**

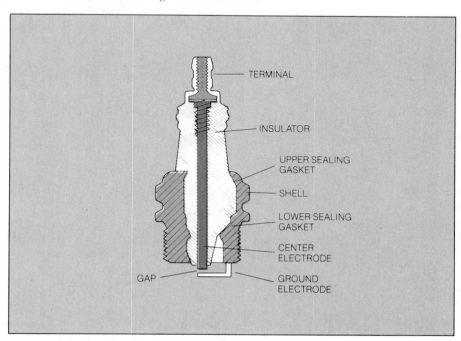

The spark is produced by applying a high voltage pulse of from 20 kV to 40 kV (1 kV is 1,000 volts) between the center electrode and ground. The actual voltage required to start the arc varies with the size of the gap, the compression ratio and the air/fuel ratio. Once the arc is started, the voltage required to sustain it is much lower because the gas mixture near the gap becomes highly ionized. (An ionized gas allows current to flow more freely.) The arc is sustained long enough to ignite the air-fuel mixture.

High Voltage Circuit and Distribution

The ignition system provides the high voltage pulse which initiates the arc. *Figure 1-9* is a schematic diagram of the electrical circuit for the ignition system. The high voltage pulse is generated by inductive discharge of a special high voltage transformer commonly called an ignition coil. The high voltage pulse is delivered to the appropriate spark plug at the correct time for ignition by a rotary switch which is called a distributor. The rotary switch is driven by the camshaft and the mechanical arrangement ensures that the high voltage is switched to the correct spark plug at the correct time. *Figure 1-10* is an illustration of a distributor.

A special type of electrical transformer, called an ignition coil, is used to create a high-voltage pulse that creates a spark at the spark plug. It is the job of the distributor to transfer this high voltage pulse to the proper spark plug at the correct time.

**Figure 1-9.
Schematic Diagram of
Ignition Circuit**

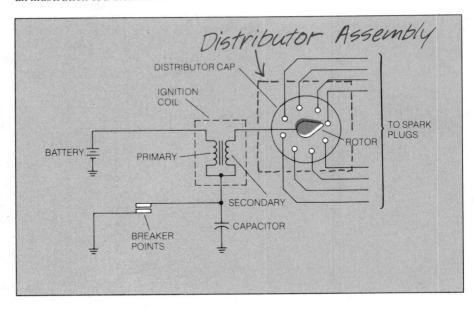

The rotor on the distributor forms a high voltage switch that distributes the voltage pulse to the selected spark plug.

A set of electrical leads, commonly called spark plug wires, are connected between the various spark plug center terminals and individual terminals in the distributor cap. The center terminal in the distributor cap is connected to the ignition coil secondary.

The rotating part of the high voltage switch, appropriately called the rotor, is driven by the camshaft and distributes the high voltage pulse from the coil to the appropriate spark plug wire. Remember that the camshaft rotates at one-half the crankshaft speed; therefore, one ignition pulse is provided to each cylinder for each two revolutions of the crankshaft.

**Figure 1-10.
Distributor**

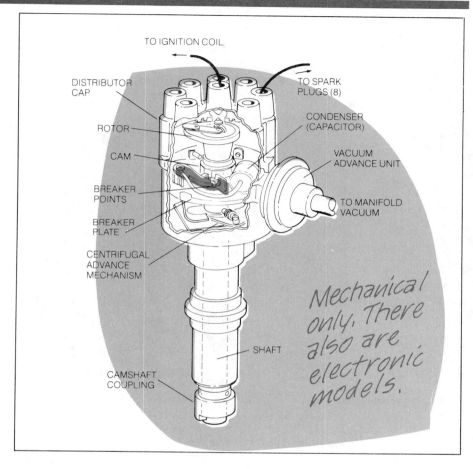

Figure contents (labels): TO IGNITION COIL, DISTRIBUTOR CAP, ROTOR, CAM, BREAKER POINTS, BREAKER PLATE, CENTRIFUGAL ADVANCE MECHANISM, CAMSHAFT COUPLING, TO SPARK PLUGS (8), CONDENSER (CAPACITOR), VACUUM ADVANCE UNIT, TO MANIFOLD VACUUM, SHAFT. Handwritten note: Mechanical only. There also are electronic models.

Primary Circuit

The distributor in a conventional ignition system uses a mechanically activated switch called breaker points. The interruption of ignition coil current when the breaker points open produces a HV pulse in the secondary.

A mechanism in the distributor of a conventional ignition system opens and closes the primary circuit of the coil by operating a switch commonly called the breaker points. During the intervals between ignition pulses (i.e., when the rotor is between contacts), the breaker points are closed (known as dwell). Current flows through the primary of the coil and a magnetic field is created which links the primary and secondary of the coil.

At the instant the spark pulse is required, the breaker points are opened (by a mechanism explained below). This interrupts the flow of current in the primary of the coil and the magnetic field collapses rapidly. The rapid collapse of the magnetic field induces the high voltage pulse in the secondary of the coil. This pulse is routed through the distributor rotor, the terminal in the distributor cap, and the spark plug wire to the appropriate spark plug. The capacitor absorbs the primary current which continues to flow during the short interval in which the points are opening and prevents arcing at the breaker points.

The waveform of the primary current is illustrated in *Figure 1-11*. The primary current increases with time after the points close (*Figure 1-11a*). At the instant the points open, this current begins to fall rapidly. It is during this rapid drop in primary current that the secondary high voltage pulse occurs (*Figure 1-11b*). The primary current oscillates (the "wavy" portion in *Figure 1-11c*) because of the resonant circuit formed between the coil and capacitor.

**Figure 1-11.
Primary Current
Waveform**

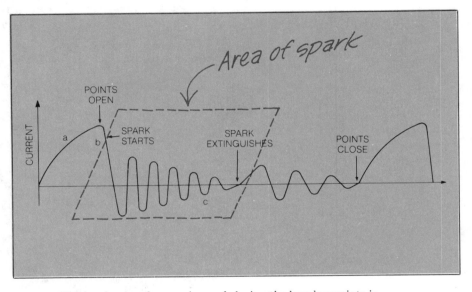

The mechanism for opening and closing the breaker points is illustrated in *Figure 1-12*. A cam having a number of lobes equal to the number of cylinders is mounted on the distributor shaft. As this cam rotates, it alternately opens and closes the breaker points. The movable arm of the breaker points has an insulated rubbing block which is pressed against the cam by a spring. When the rubbing block is aligned with a flat surface on the cam, the points are closed as shown in *Figure 1-12a*. As the cam rotates, the rubbing block is moved by the lobe (high point) on the cam as shown in *Figure 1-12b*. At this time, the breaker points open and spark occurs.

A multisurfaced cam, mounted on the distributor shaft, is used to open and close the breaker points (Figure 1-12).

Ignition Timing

The point at which ignition occurs, in comparison to the top dead center of the piston's compression stroke, is known as ignition timing.

Ignition occurs some time before top dead center (BTDC) during the compression stroke of the piston. This time is measured in degrees of crankshaft rotation BTDC. For a modern SI engine, this timing is typically 8 to 10 degrees for the basic mechanical setting with the engine running at low speed (RPM). This basic timing is set by the design of the mechanical coupling between crankshaft and the distributor. The basic timing may be adjusted slightly in many cars by physically rotating the distributor housing.

As the engine speed increases, the angle through which the crankshaft rotates in the time required to burn the fuel and air mixture increases. For this reason, the spark must occur at a larger angle BTDC for higher engine speeds. This change in ignition timing is called spark advance.

Figure 1-12.
Breaker-Point Operation

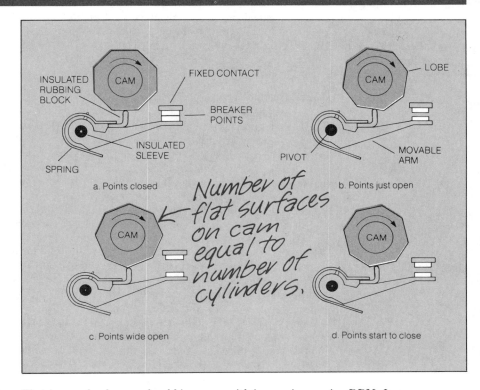

INSULATED RUBBING BLOCK
CAM
FIXED CONTACT
BREAKER POINTS
INSULATED SLEEVE
SPRING

a. Points closed

LOBE
CAM
PIVOT
MOVABLE ARM

b. Points just open

Number of flat surfaces on cam equal to number of cylinders.

CAM

c. Points wide open

CAM

d. Points start to close

Spark advance changes ignition timing as the speed of an engine increases. A set of levers and weights within the distributor provides it in the SI engine.

Ignition timing also varies with atmospheric pressure within the intake manifold. A vacuum advance mechanism senses the pressure within the intake manifold, and adjusts the timing accordingly.

That is, spark advance should increase with increasing engine RPM. In a conventional ignition system, the mechanism for this is called a centrifugal spark advance. It is shown in *Figure 1-10*. As engine speed increases, the distributor shaft rotates faster, and the weights are thrown outward by centrifugal force. The weights operate through a mechanical lever so their movement causes a change the relative angular position between the rubbing block on the breaker points and the distributor cam and advances the time when the lobe opens the points.

In addition to speed dependent spark advance, there is a need to adjust the ignition timing as a function of intake manifold pressure. Whenever the throttle is nearly closed, the manifold pressure is low (i.e., nearly a vacuum). The combustion time for the air-fuel mixture is longer for low manifold pressure conditions than for high manifold pressure conditions (i.e., near atmospheric pressure). As a result, the spark timing must be advanced for low pressure conditions to maintain maximum power and fuel economy. The mechanism to do this is a vacuum operated spark advance. This is also shown in *Figure 1-10*. The vacuum advance mechanism has a flexible diaphragm connected through a rod to the plate on which the breaker points are mounted. One side of the diaphragm is open to atmospheric pressure and the other side is connected through a hose to manifold vacuum. As manifold vacuum increases, the diaphragm is deflected (atmospheric pressure pushes it) and moves the breaker point plate to advance the timing.

Because ignition timing is critical to engine performance, controlling it electronically is a major automotive application.

Ignition timing significantly affects engine performance and exhaust emissions; therefore, it is one of the major factors that is electronically controlled in the modern SI engine. The performance of the ignition system and the spark advance mechanism has been greatly improved by electronic control systems.

DIESEL ENGINE

Clearly the vast majority of automobile engines are SI engines. For many years, alternative engines such as the diesel were simply not able to compete effectively with the SI engine in the United States. Diesel engines were used mostly in heavy duty vehicles such as large trucks, ships, railroad locomotives, earthmoving machinery, etc. The use of diesel engines in passenger cars has been slow in coming to the U.S., although now its use is increasing rapidly. The motivation for this increased application comes largely from the lower cost of diesel fuel as compared to gasoline and from the better fuel economy of the diesel engine as compared to the gasoline engine.

Diesel engines are similar in basic design to SI engines. However, diesel engines lack an ignition system. Instead, the diesel engine relies on a combination of fuel injection and very high pressure to produce the ignition sequence required for combustion.

In many ways, the 4 stroke/cycle diesel engine closely resembles the SI engine. It has a crankshaft, cylinders, pistons, intake and exhaust valves and is usually water cooled. However, there are significant differences between these two classes of engine. For example, the diesel engine has no ignition system for normal engine operation. Moreover, the cylinder head, in the best case, contains a fuel injector for each cylinder instead of a spark plug. (A glow plug is used to help heat the air only during starting.) Instead of a high voltage distribution system there is a fuel distribution system. It consists of a fuel pump, fuel filters, a mixture controller, fuel distribution lines, and fuel injectors. *Figure 1-13* shows a cutaway view of a passenger car diesel engine.

The diesel engine has a higher compression ratio and a heavier, stronger construction than the SI engine. The modern SI engine's compression ratio is typically around 8:1 and rarely exceeds 12:1. The diesel engine's compression ratio is typically around 21:1. The cylinder pressure during the power stroke of a diesel has a very rapid rise (almost like an explosion) which places a high stress on the engine's structure. This is the reason for the heavy construction.

The theory of operation of the diesel engine can be explained with reference to *Figure 1-14*. In the diesel engine, the fuel and air are not mixed external to the cylinder as they are in the gasoline engine. After an engine is running and during the intake stroke, the intake valve is open and air alone is drawn into the cylinder. During the compression stroke, this air is compressed by a very large ratio to a very high cylinder pressure of about 500 psi. This high compression heats the air to about 1,000 degrees F. (A simple way to realize the heating effect of compressing air is to feel the bottom of a bicycle air pump after it has been used for a few minutes to inflate tires). When the fuel is injected near the top of the compression stroke, it is ignited by this high

**Figure 1-13.
Four-Cyiinder Diesel
Engine for Passenger
Cars**

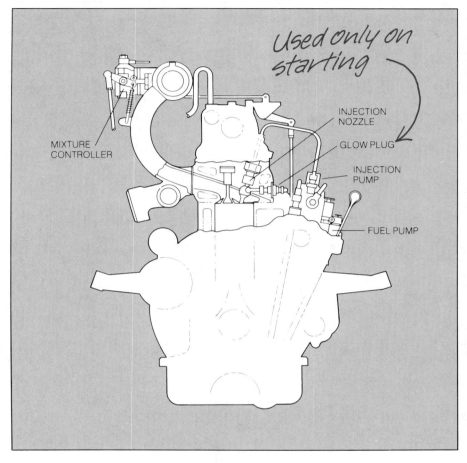

pressure, high temperature compressed air. The cylinder pressure rises
rapidly and the tremendous force drives the piston down during the power
stroke. Finally, on the exhaust stroke, the exhaust valve opens and the
combustion products are forced out of the cylinder. This four-stroke cycle
repeats continuously.

One of the important parameters of the diesel engine is the timing of
the fuel injection relative to piston position. (This corresponds to the spark
timing in the SI engine.) The power produced by the engine is significantly
affected by this timing. The control of fuel injection timing is a potential
application for an electronic control system.

Another parameter affecting diesel engine performance is the speed
with which the fuel is injected into the cylinder. The time interval of the
combustion process is influenced by this injection speed. There is also a
potential for electronically controlling this speed to further optimize engine
performance. However, since the majority of automotive engines are the SI
engine, the remainder of the book is concerned only with the SI engine.

The timing of the fuel injection is critical to a diesel engine, just as the
timing of the spark is critical to a SI engine.

Figure 1-14.
Four-Stroke Diesel Cycle

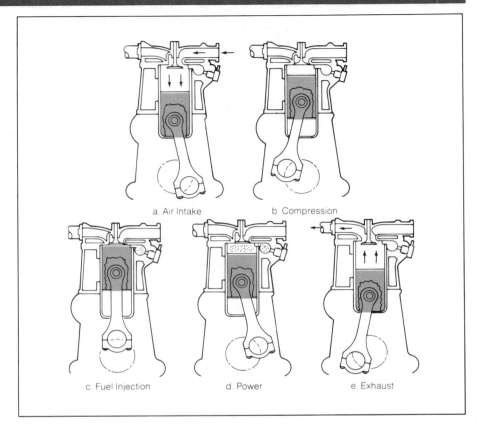

a. Air Intake   b. Compression

c. Fuel Injection   d. Power   e. Exhaust

DRIVE TRAIN

The engine drive train system of the automobile consists of the engine, transmission, driveshaft, differential and driven wheels. We have already discussed the SI engine and we know that it provides the motive power for the automobile. Now let's examine the transmission, driveshaft, and differential in order to understand the role of these devices.

Transmission

The transmission provides a match between engine speed, which should remain relatively constant, and vehicle speed, which varies greatly.

The transmission is a gear system which adjusts the ratio of engine speed to wheel speed. Essentially, the transmission enables the engine to operate within its optimum performance range, regardless of the vehicle load or speed. It provides a gear ratio between the engine speed and vehicle speed such that the engine provides adequate power to drive the vehicle at any speed.

To accomplish this with a manual transmission, the driver selects the correct gear ratio from a set of possible gear ratios (usually 3 to 5 for passenger cars). An automatic transmission selects this gear ratio by means of an automatic control system. Most automatic transmissions have three forward gear ratios, although a few have two and some have four. A properly used manual transmission normally has efficiency advantages over an automatic transmission, but the automatic transmission is the most used transmission for passenger automobiles in the U.S. The present day automatic transmissions are controlled by a hydraulic and pneumatic system, but there is a possibility for electronic controls to be used in the future. The control system must determine the correct gear ratio by sensing the driver select command, accelerator pedal position, and engine load.

Driveshaft

The driveshaft is used on front engine, rear wheel drive vehicles to couple the transmission output shaft to the differential input shaft. Flexible couplings, called universal joints, allow the rear axle housing and wheels to move up and down while the transmission remains stationary.

Differential

The combination of driveshaft and differential complete the transfer of power from the engine to the rear wheels.

The differential serves three purposes. The most obvious is the right angle transfer of the rotary motion of the driveshaft to the wheels. The second purpose is to allow each driven wheel to turn at a different speed. This is necessary because the "outside" wheel must turn faster than the "inside" wheel when the vehicle is turning a corner. The third purpose is the torque increase provided by the gear ratio. This gear ratio can be changed to allow different torque to be delivered to the wheels while using the same engine and transmission. This gear ratio also affects fuel economy.

SUSPENSION, STEERING AND BRAKES

Instead of direct attachment of the axles and wheels to the frame, they are isolated from the frame by a suspension system to provide a ride inside the car that is much smoother than the road surface over which the car is being driven.

The steering mechanism attached to the front wheels permits the driver to control the direction of vehicle motion by turning the steering wheel.

The drum or disk brakes installed on the wheels are the means used to bring the automobile to a stop once it is in motion.

Some electronic controls have been applied to braking systems, but thus far, electronic control has not been practical for the suspension and steering of the automobile.

SUMMARY

In this chapter, we have briefly reviewed the major systems of the automobile and discussed the basic engine operation. In addition, we have indicated where electronic technology could be applied to improve performance or reduce cost.

The next few chapters of this book are intended to develop a basic understanding of electronic technology. Then we'll use all this knowledge to examine how electronics has been applied to the major systems. In the last chapter, we'll look at some ideas and methods that may be used in the future.

Quiz for Chapter 1

1. The term TDC refers to
  a. the engine exhaust system.
  b. rolling resistance of tires.
  c. crankshaft position corresponding to a piston at the top of its stroke.
  d. the distance between headlights.

2. The distributor is
  a. a rotary switch which connects the ignition coil to the various spark plugs.
  b. a system for smoothing tire load.
  c. a system which generates the spark in the cylinders.
  d. a section of the drivetrain.

3. The air/fuel ratio is
  a. the rate at which combustible products enter the engine.
  b. the ratio of the mass of air to the mass of fuel in a cylinder before ignition.
  c. the ratio of gasoline to air in the exhaust pipe.
  d. intake air and fuel velocity ratio.

4. Ignition normally occurs
  a. at BDC.
  b. at TDC.
  c. just after TDC.
  d. just before TDC.

5. Most automobile engines are
  a. large and heavy.
  b. gasoline fueled, spark ignited, liquid cooled internal combustion type.
  c. unable to run at elevations which are below sea level.
  d. able to operate with any fuel other than gasoline.

6. An exhaust valve is
  a. a hole in the cylinder head.
  b. a mechanism for releasing the combustion products from the cylinder.
  c. the pipe connecting the engine to the muffler.
  d. a small opening at the bottom of a piston.

7. Power is produced during:
  a. intake stroke.
  b. compression stroke.
  c. power stroke.
  d. exhaust stroke.

8. The transmission
  a. converts rotary to linear motion.
  b. optimizes the transfer of engine power to the drivetrain.
  c. has 4 forward speeds and one reverse.
  d. automatically selects the highest gear ratio.

9. The suspension system
  a. partially isolates the body of a car from road vibrations.
  b. holds the wheels on the axles.
  c. suspends the driver and passengers.
  d. consists of 4 springs.

10. The camshaft
  a. operates the intake and exhaust valves.
  b. rotates at the same speed as the crankshaft.
  c. has connecting rods attached to it.
  d. opens and closes the breaker points.

11. An SI engine is:
  a. a type of internal combustion engine
  b. a Stirling engine
  c. always fuel injected
  d. none of the above

12. The intake system refers to:
  a. the carburetor
  b. a set of tubes
  c. a system of valves, pipes and throttle plates
  d. the components of an engine through which fuel and air are supplied to the engine

The Systems Approach to Control and Instrumentation

ABOUT THIS CHAPTER

Modern day scientists know that the physical laws and processes interact with one another and that it is necessary to consider them together as a system rather than individually. Consequently, the theory of systems has been well developed and is used extensively in applied science and engineering.

This chapter is about systems and the systems approach to control and instrumentation. Topics of discussion include how to describe what a system does through diagrams (qualitatively) and how well a system performs its function (quantitatively) through the use of models. The basics of control and instrumentation are presented and the systems approach is applied to some automotive examples. The material in this chapter lays a foundation for understanding the automotive systems in later chapters.

SYSTEMS

What image does the word system bring to your mind? Perhaps the solar system of which our earth is a part. Examples of systems are easy to find, yet they generally don't have much in common. For instance, the bones in our body make up a system called a skeleton which supports muscles, arteries, veins, and organs which are part of our cardiovascular and nervous systems. The way we live and conduct ourselves is a social system. The exchange of goods and services is an economic system. We listen to a stereo system and we live within an ecological system. So how can a word (system) be defined that can be applied to such variation? Notice that although each of these systems is unique, they all have at least one thing in common—*each system is a collection of interacting parts*.

All systems share a common trait. They are made up of a combination of parts that interact with one another.

However, this raises another problem—the parts of a system are sometimes difficult to pinpoint. The skeletal system is made up of joints and bones which are easily identified. An economic system, on the other hand, is made up of producing units, consuming units, government units, and a variety of other units which interact to make the economy run in ways that sometimes are not obvious.

Systems can often be broken down into a number of subsystems. The subsystems also consist of a number of individual parts.

A stereo system consists of a phonograph, amplifier, and loudspeakers as shown in *Figure 2-1a*, but a closer look reveals the phonograph has a number of parts of its own such as a cartridge and needle, a platter and motor, and a tone arm and switches. The amplifier has even more parts. It is apparent that the stereo system is made up of parts which themselves can be considered a system while being a subsystem (part) of the stereo system. Each of these subsystems can be further divided into other subsystems all the way down to the individual nuts and bolts which make up the lowest level of the system.

**Figure 2-1.
System Diagrams**

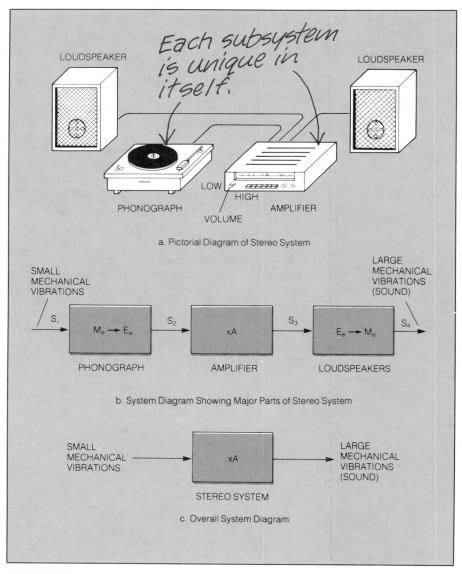

a. Pictorial Diagram of Stereo System

b. System Diagram Showing Major Parts of Stereo System

c. Overall System Diagram

Purpose of System Analysis

A functional system analysis is called qualitative analysis, and such an analysis is based around the qualities of that system.

The analysis of a system is aimed toward determining either the function or performance of the system. *A functional analysis is known as a qualitative analysis* because the goal of the analysis is to determine certain qualities of the system. A qualitative analysis answers the question: what does a system do? *A performance analysis is called a quantitative analysis.* A quantitative analysis attempts to measure one or more quantities of a system which affect how well a system performs its function.

A system performance analysis is called quantitative analysis. It is based around measurement of the performance of a given system.

Usually a complete analysis will include both a qualitative and quantitative analysis. They complement each other and aid in the overall understanding of the system's operation. The qualitative analysis sets the groundwork and builds the frame, then the quantitative analysis completes the study by filling in the details. A qualitative analysis usually takes the form of a system diagram showing all of the important parts along with a discussion of the function of each part. A quantitative analysis determines the performance of each component and shows how that performance affects the performance of the entire system. Performance is evaluated by measuring key parameters that affect the system.

QUALITATIVE ANALYSIS

A qualitative analysis uses all given knowledge about a system's major parts to determine the systems' function. In this process of analysis, unnecessary components are not considered.

A qualitative analysis begins by using all the available knowledge about a system's parts to eliminate all unnecessary components and subsystems from consideration. The function of a system is determined by the function of its parts, so the next step is to determine the function of the system's parts.

In the stereo system example of *Figure 2-1*, the operation of the first level subsystems is rather obvious so no further investigations into the functions of the lower levels is necessary. The phonograph is known to convert (transform) the mechanical energy produced by dragging a needle across irregularities in the grooves of a recording into electrical energy. The electrical energy is transferred from the phonograph to the amplifier which increases the power of the signal by an amount determined by the volume control on the amplifier. The amplifier then sends the amplified signal to the loudspeakers which convert the electrical energy into mechanical energy in the form of sound vibrations.

System Diagram

A system diagram should illustrate a system's function in an efficient and straightforward manner. *Figure 2-1a* is a good representation of a stereo system's appearance, but doesn't tell the viewer what the system does (assuming the viewer is not already familiar with such a system). Therefore, it is an inadequate representation of the system for qualitative analysis. *Figure 2-1a* shows the needle, tone arm, platter, and even the on-off switch of the phonograph, but doesn't convey the important function of the device.

Figure 2-1b is a better diagram for system analysis because it describes in more detail the function of each component. The phonograph takes mechanical vibrations, signal S_1, and transforms the mechanical energy M_e of S_1 into electrical energy E_e. This comes out of the phonograph as signal S_2 and enters the amplifier where it is multiplied by a variable gain A to produce an amplified signal S_3. S_3 is transmitted to the loudspeakers where it is transformed from electrical energy to mechnical energy sound vibrations, signal S_4. *Figure 2-1b* reveals the overall effect of the stereo system as a system that amplifies small mechanical vibrations, S_1, into large mechanical vibrations, S_4, which produce the sound. Signal S_2 and S_3 are intermediate signals and are used only to identify the electrical nature of the amplifier. Intermediate signals help to describe how a system performs its function.

Now a simpler diagram, *Figure 2-1c*, can be used to show the overall function of the system. Since it excludes much of the detail, the electrical nature of the amplifier is not shown. However, if the overall function of the system is all that is desired, *Figure 2-1c* is really all that is needed. It shows small mechanical vibrations being amplified into larger mechanical vibrations.

Block Diagrams

Block diagrams are commonly used to represent each major component or subsystem within a system, and can quickly show important relationships inside the system.

The system diagrams in *Figures 2-1b* and *2-1c* are known as block diagrams. Each major component or subsystem is represented by a labeled block and lines with arrowheads are drawn between blocks to show signal flow direction. Block diagrams are useful because they allow the viewer to learn the important systems information quickly and efficiently. It takes a lot of time and artistic skill to carefully draw a recognizable picture of a system component such as a phonograph, but it requires only a little time and effort to draw a box and label it appropriately. Block diagrams are used quite often by people who design and use systems, so they will be used in this book when we discuss automotive systems.

System designers have developed some standard system symbols. *Figure 2-2* shows four standard system symbols; summing block, multiplier, gain block, and take-off point.

QUANTITATIVE ANALYSIS

Quantitative analysis measures a system's performance by comparing that performance to other systems.

The goal of a quantitative analysis is to determine how well a system performs its function. To do this, a means must be developed so a system's performance may be evaluated and compared fairly to the performance of other systems. Many times this is a challenging task because it is not always easy to pinpoint the performance of a system. For example, when trying to determine how well an amplifier amplifies, one must take into consideration some of the specific features of the amplifier. The function of an amplifier is to increase the power level of an input signal. In so doing, a practical amplifier produces a certain amount of noise which is added to the desired signal. In addition, an amplifier may distort the input signal. As a result, the amplified output waveform looks slightly different than the input waveform. (For an audio amplifier that might be part of a stereo system, the noise and distortion cause unpleasant sounds.) These characteristics of the amplifier change with the amount of amplification, the frequency of the input signal, and other conditions.

**Figure 2-2.
Standard System
Symbols**

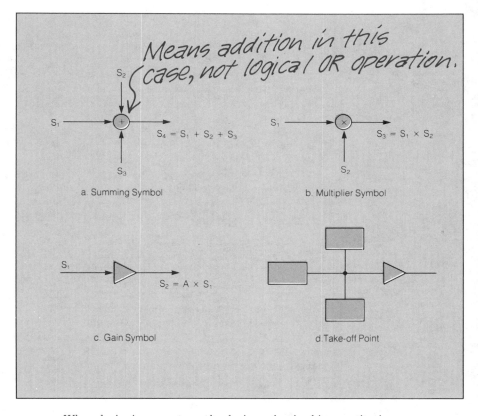

a. Summing Symbol

b. Multiplier Symbol

c. Gain Symbol

d. Take-off Point

When designing a system, the designer begins his quantitative analysis with a knowledge of which parameters affect the performance of the system and attempts to choose system parts which will provide acceptable performance. This requires that the designer have a detailed knowledge of the operation of each subsystem he uses. The designer obtains this knowledge from the qualitative analysis as described in the preceding section.

The performance of each system part is generally described by a mathematical equation which relates the part's output to its input. The part, then, is known to the designer by its mathematical equation and is compared with other similar parts by comparing equations. This may seem like a rather abstract way of comparing system parts, but it is actually quite useful because the mathematical equations are not very sensitive to a person's opinions or feelings. Therefore, the results of a comparison are more accurate.

The performance of the parts within a system can be described in terms of mathematical equations. These equations can then be compared to equations that describe other similar parts.

SYSTEM MODELING

A mathematical equation that is used to represent part or all of a system is called a model.

A mathematical relation used to represent a system part, subsystem or total system is known as a mathematical model, or just model. Mathematical models are used extensively in the design phase of system development. The mathematical model can be programmed on a computer to simulate the performance of a system and the model can be easily changed by changing inputs to the computer. If actual parts were used, changes in design would result in changing components and/or connections. This is a time consuming and costly process.

Models are usually developed from knowledge of how a system or part has performed its function in other systems. This available knowledge, along with a knowledge of physical laws, is used as the basis for a mathematical equation that describes the part's function. This is not an easy task and may take a great deal of effort to develop. However, once proven, it is a very useful tool.

There are times when the designer doesn't know how a system part actually performs its function, but does know how it responds to various input signals and what outputs are produced. This type of system part is typically called a "black box". Particular mathematical equations may be assigned to these "black boxes" without actually having a physical part to accomplish the function.

Frequency Response

A frequency response plot for an audio amplifier is one kind of mathematical model. It quantitatively describes the amplifier's performance.

A system's frequency response is one type of mathematical model. As mentioned earlier, a frequency response quantitatively describes an audio amplifier's performance. Many readers may have seen a frequency response plot (graph of amplitude vs frequency) among the literature at their favorite audio store. *Figure 2-3* shows a typical linear frequency response plot of a high fidelity amplifier. The frequency response shows that the amplifier amplifies quite well (by a factor of about 100) all signals within a frequency band between about 100 and 18,000 hertz. It amplifies poorly (with a low gain) signals lower than 50 hertz or higher than 19,000 hertz. Thus, a frequency response plot provides important information to the audio amplifier buyer because it provides one indicator of how well the amplifier will reproduce its input.

Notice in the frequency response of *Figure 2-3* how the low frequency part of the plot, between 0 and 50 hertz, is packed tightly together in the first part of the plot while the high frequency portion, between 10,000 and 20,000 hertz, takes about half of the plot. Many times the low frequency part of the plot has information which cannot be seen when it's squeezed together. The Bode (Bō-dē) plot of *Figure 2-4* has the same information as *Figure 2-3*, but spreads it more evenly by using a logarithmic frequency scale and decibel magnitude scale.

Figure 2-3.
Audio Amplifier
Frequency Response
Linear Plot

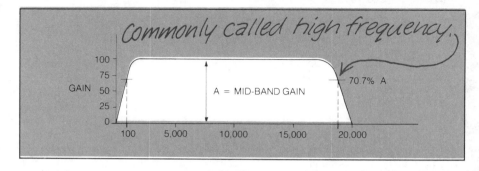

Figure 2-4.
Audio Amplifier
Frequency Response
Bode Plot

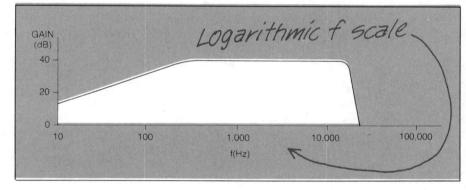

The value of a decibel is given by the following equation:

$$dB = 20 \log_{10} A$$

It means that one decibel is equal to twenty times the logarithm to the base ten of the gain A. Remember that the logarithm of a number is the exponent to which the base must be raised to get the number. If the gain of an amplifier is 100, then the base 10 must be raised to the second power (have an exponent of 2) to get 100. Therefore, a gain of 100 in decibels is:

$$dB = 20 \log_{10} 100$$
$$dB = 20 \times 2$$
$$dB = 40$$

Table 2-1 summarizes the conversion to decibels for some common gains.

Gain	Decibels
1	0
10	20
100	40
1000	60

Frequency plots that have logarithmic frequency scales are normally used because they are easier to read. Amplitudes may be on a linear or decibel scale.

The frequency response plot can be used to describe the input-output relation for many systems. It becomes one of the system designer's and system user's most useful tools. One of the ways that a "black box" in a system can be modeled is by recording its response to a variety of different inputs and determining a system model which is compatible with each.

The frequency response plot of a system is a convenient way of recording the system's response to an important class of inputs called sinusoids. *Figure 2-5* shows different sinusoidal signals being input to a system and the resulting output from the system. The frequency response plot *(Figure 2-6a)* is generated by recording and plotting the ratio of the output signal's amplitude to the input signal's amplitude (i.e., the system gain). The phase difference between the input and output is also recorded and plotted as shown in *Figure 2-6b*. The frequency response and phase plot enable the designer and user to determine how the system will respond to an input sinusoid of a certain frequency. The amplitude plot shows how much the signal is amplified and the phase plot shows how much the output is delayed in time as it passes through the system.

The system's response to non-sinusoidal signals can be determined by using a mathematical technique developed by Joseph Fourier. *Figure 2-7* shows how a non-sinusoidal signal (a square wave) can be approximated by adding several specifically selected sinusoidal signals called the Fourier series. The response to the square wave can be determined by inputting each sinusoid separately as in *Figure 2-5* and adding all of the system's computed responses.

A frequency response plot of a particular system can enable the systems' designer or user to determine how the system will react to an incoming signal of a certain frequency.

A system designer can predict or compute a system's response to many different types of inputs by using the frequency response plot. This makes it a very powerful tool and it will be used to help describe various automotive systems in later chapters.

Transfer Functions

To model a system, a mathematical equation is required to represent the system function. The mathematical equation whose plot is the same as the frequency response plot of a system or a system's part is called a transfer function. Instead of looking at a graph with the gain and phase shift of the system plotted, the equation can be used to determine the gain and the phase shift. The transfer function will show that the input signal is amplified by the product of the gains of the individual subsystems and the phase is shifted by the sum of the phase shifts of the individual subsystems.

**Figure 2-5.
System Response to
Different Sinusoid Inputs**

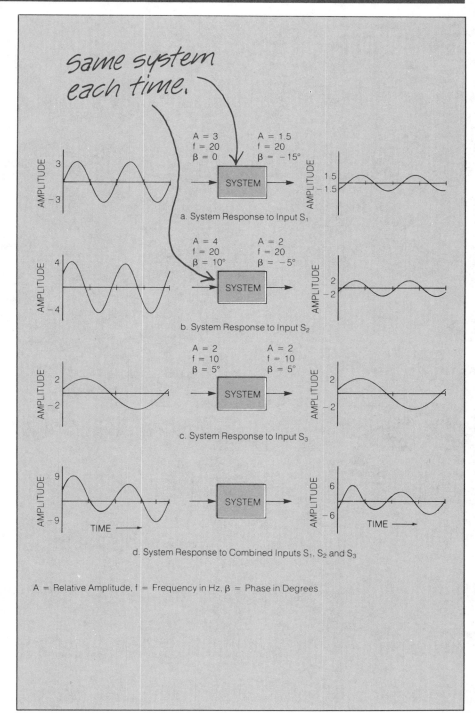

a. System Response to Input S_1

b. System Response to Input S_2

c. System Response to Input S_3

d. System Response to Combined Inputs S_1, S_2 and S_3

A = Relative Amplitude, f = Frequency in Hz, β = Phase in Degrees

Figure 2-6.
Frequency Response
and Phase Shift Plots of
System in Figure 2-5

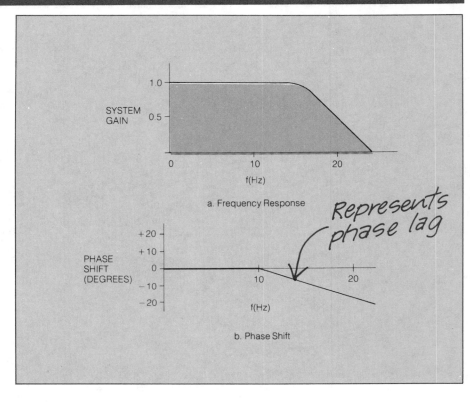

a. Frequency Response

Represents phase lag

b. Phase Shift

INSTRUMENTATION SYSTEMS

Instrumentation systems,
whether they are elec-
trical, mechanical, or a
combination of both, mea-
sure a physical quantity
and provide a report of
that measurement.

Instrumentation systems measure a physical quantity and report the
measurement in some convenient form. Examples of automotive
instrumentation systems include:

1) Speedometers
2) Odometers
3) Tachometers
4) Fuel gauges
5) Oil pressure gauges
6) Clocks
7) Engine parameter warning indicators
8) Door ajar indicators
9) Ammeters

Some of these systems are mechanical, some are electrical, and some are
combinations of electrical and mechanical subsystems. It doesn't matter how
an instrumentation system is made and used, the general goal of the
instrumentation system is universally the same—to provide an accurate,
reliable and readable measurement of a particular physical quantity.

Figure 2-7.
A Square Wave Can Be
Represented by a Fourier
Series of Sinusoids

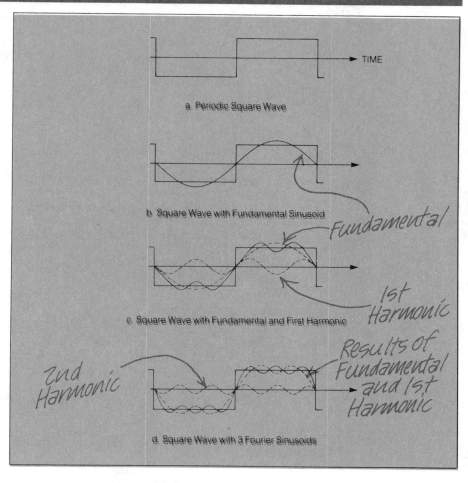

a. Periodic Square Wave

b. Square Wave with Fundamental Sinusoid

Fundamental

c. Square Wave with Fundamental and First Harmonic

1st Harmonic

2nd Harmonic

Results of Fundamental and 1st Harmonic

d. Square Wave with 3 Fourier Sinusoids

Accuracy

The accuracy of an instrument describes how close the instrument's measurement and indication of a physical quantity is to the actual quantity. Differences between actual and measured values arise from two important sources of errors; systematic errors and random errors.

Errors in the accuracy of instruments are due to systematic errors, caused by defects within the system, and by random errors, caused by outside disturbances.

Systematic errors affect an instrument's basic accuracy. Systematic errors are errors due to such things as improper calibration, temperature drift, or improper use of the instrument. These can be corrected for or eliminated by the operator.

Random errors are caused by electrical, mechanical, or any other type of noise disturbance which adversely affects the measurement of the physical quantity. These errors can be eliminated by filtering, shielding or selecting system parts with low internal noise.

Reliability

The reliability of an instrumentation system refers to its ability to perform its designed function accurately and continuously whenever required, under unfavorable conditions, and for a reasonable lifetime. Reliability must be designed into the system by using adequate design margins and quality components that operate over the desired temperature range and other environmental conditions.

Readability

The readability of an instrument describes how easily the results of a measurement can be read and understood by a human. A properly displayed measurement should indicate the numerical value of the measurement along with the units of the measured quantity. This is called calibration. For example, on a speed indicator dial, it is necessary to have not only the major divisions marked with miles per hour and kilometers per hour, but subdivisions that quickly make sense to the reader at a glance.

Readability also depends on the size of the display, its location in relationship to its surroundings, the angle of viewing, how well it is lighted, and the contrast of the display to its background. Color can have a very important impact.

Basic Measurement System

An instrumentation system consists of three basic parts; a sensor, a signal processing unit, and a form of display device or actuator device.

A basic measurement system is depicted in *Figure 2-8*. This is represented as an electrical system model. The basic system could be electrical, mechanical, pneumatic, or a combination. Notice that this diagram is a generalized form of *Figure 2-1b*. An instrumentation system is comprised of three basic parts: 1) a sensor which converts the physical quantity q_0 into an electrical signal q_1 so that it may be operated on by the signal processor; 2) a signal processor which performs some operation on the intermediate signal q_1 to increase its power level, reliability, and accuracy and to put it into a form so that when displayed, it can be understood by humans; and 3) a display device

**Figure 2-8.
Basic Measurement
System Diagram**

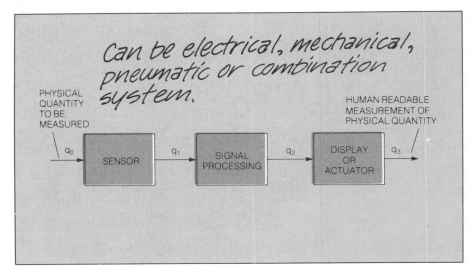

or actuator which converts the signal q_2 from the signal processor into some readable quantity or some action q_3. The ratio of q_0 to q_3 should be 1 for a properly calibrated instrument. The calibration can be changed by altering any of the 3 parts. Each part will be discussed separately and the role it plays in the operation of the entire instrumentation system will be presented.

Sensors

Sensors convert one form of energy, such as thermal energy, into another form of energy (often electrical).

A sensor is a device that converts energy from one form to another. Ideally, it is sensitive only to the effects of a particular physical quantity. Typical automotive quantities are speed, torque, air pressure, temperature, fuel flow rate, throttle angle, and air/fuel ratio. Sensors convert a portion of the energy of the quantity to be measured into a form which can be used by the signal processor.

Sensors are designed to operate over a given range of input values of the physical quantity they are measuring. This is known as the sensor's dynamic range. The typical dynamic range for an automotive coolant temperature sensor is from 100°F to 300°F.

Another sensor parameter is the ratio of output value to input quantity. This is called the sensor's calibration coefficient which, for a linear static sensor, is the slope of its transfer characteristic. For a throttle angle sensor which converts a throttle angle of 0° to 90° to a voltage between 0 and 5 volts, the calibration coefficient is given by:

$$\frac{(5V - 0V)}{(90° - 0°)} = \frac{1V}{18°}$$

or about 0.056 volt per degree.

Sensors with a wide bandwidth and good linearity accurately respond to a wide range of varying conditions.

Systematic errors in sensors are commonly due to sensor bandwidth and linearity. Ideal dynamic sensors display a flat frequency response (the gain is constant over a wide frequency range as shown in the midrange of *Figure 2-3*) so the sensor doesn't distort the amplitude of different frequency components of the input signal. The *bandwidth of a sensor* (or system) is the frequency at which the transfer function ceases being flat and the gain of the sensor drops below some predetermined minimum value. As shown in *Figure 2-3*, for common frequency response plots, the bandwidth end points are determined where the response has dropped to 70.7% of its mid-band response. If the response is plotted in decibels, the end points are where the response has dropped 3dB from its mid-band value.

Because of the complexities of a sensor in changing energy from one form to another, the sensor output can be in error compared to the input.

Real sensors many times have a complicated transfer function; thus, errors exist between the input and the output signals. *Figure 2-9* shows the input and output of a sensor whose bandwidth causes distortion of the output. Notice the sharp corners of the input are rounded by the sensor. Sharp corners represent high frequencies in the signal's Fourier series; therefore, the rounded corners indicate that the sensor's bandwidth is not wide enough. It's gain at high frequencies is small compared to its gain at lower frequencies.

**Figure 2-9.
Distortion Caused by
Sensor**

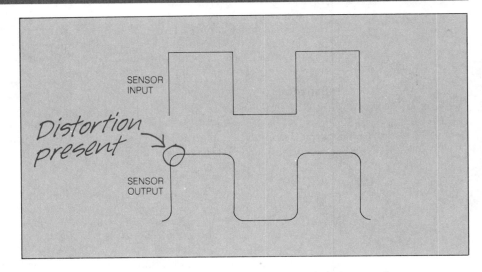

Signal processing can be
used to compensate for
systematic errors of
sensors.

An ideal sensor has a linear transfer characteristic (or transfer function) as shown in *Figure 2-10a* to allow the resulting measurements to be easily interpreted. More often, however, the transfer characteristic of real sensors is nonlinear as shown in *Figure 2-10b*. Thus, some signal processing is required to linearize the output signal so it will appear as if the sensor had a straight line (linear) transfer characteristic as shown in the dashed curve of *Figure 2-10b*. Sometimes a nonlinear sensor may provide satisfactory operation without linearization if it is operated in a particular linear region of its transfer characteristic as shown in *Figure 2-10b*.

**Figure 2-10.
Sensor Transfer
Characteristics**

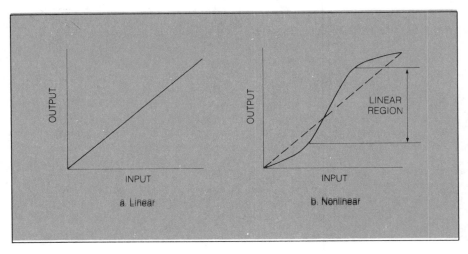

Sensors are subject to random errors such as heat, electrical noise, and vibrations.

Random errors in electronic sensors are caused by internal electrical noise and imperfect selectivity. Internal electrical noise can be caused by molecular vibrations due to heat (thermal noise), random electron movement in semiconductors (shot noise), and radioactive decay. Imperfect selectivity is the sensor's inability to screen out the effects of other quantities which occur along with the quantity to be measured. For example, the output of a sensor that is measuring pressure may also change as a result of temperature changes. An ideal sensor will respond only to one physical quantity or stimuli. However, real sensors are rarely, if ever, perfect and will generally respond in some way to outside disturbances.

Displays and Actuators

Automotive display devices, typically analog or digital meters, provide a visual indication of the measurements made by the sensors.

An instrumentation system must somehow make its measurements available to the user. This is done through a display device or an actuator. A typical automotive display device is the analog meter or gauge that uses an electromagnetic force to drive an indicating pointer across a dial with appropriate markings. This device is used for speedometers, tachometers, and fuel level, oil pressure, and battery voltage gauges. Liquid crystal displays and light-emitting diode displays (similar to those used in watches) and indicator lights are also used in automotive instrumentation to report measurement results.

Actuators are subject to the same type errors in their response as sensors.

Automotive actuators include vacuum controlled diaphragms and switches and solenoid controlled valves and switches. These are used for throttle positioners for cruise control, spark timing advance mechanisms, and fuel metering valves.

Displays and actuators, like sensors, are energy conversion devices and have the same attributes and faults as sensors. Display devices have bandwidth, dynamic range, and calibration characteristics. They have the same types of errors as sensors. Many shortcomings of actuators, displays, and sensors can be reduced or eliminated through the imaginative use of signal processing.

Signal Processing

Any changes performed on the signals between the sensor and the display or actuator is considered to be signal processing.

Signal processing, as defined earlier, is any operation performed on the intermediate signals between sensor and display or actuator. It couples the sensor to the display or actuator, and increases the accuracy, reliability, or readability of the measurement. Signal processing may be used to make a nonlinear sensor or actuator appear linear, or it may smooth a sensor's or actuator's frequency response. Signal processing may be used to perform unit conversions such as converting from miles per hour to kilometers per hour; display formatting such as scaling and shifting a temperature sensor's output so that it can be displayed on the engine temperature gauge from 100° to 300°F instead of from 0° to 200°F; or process the signals in a way that reduces the effects of random system errors.

Signal processing can use analog circuitry or digital circuitry.

Signal processing can be done with analog devices or digital devices. Analog signal processing uses amplifiers, filters, adders, multipliers and other analog components. Digital signal processing uses logic gates, counters, binary adders, microcomputers, and other digital components. The difference between analog signals and digital signals is that analog signals are continuously variable and can be any value within a range while a digital signal changes in discrete steps and can take only certain values within a range. A speedometer indicates an automobile's speed using an analog signal which can have any value between zero miles per hour and its maximum speed. A digital signal from a door switch is used to indicate whether a door is open or closed. It has two values, either 0 volts or +12 volts; 0 for closed, 12 for open. Chapter 4 will have more about digital signal processing and present some specific hardware used to convert analog signals to digital and digital to analog.

CONTROL SYSTEMS

Control systems are systems that are used to direct the operation of other systems. For our discussion, the system being controlled is known as the *system plant*. The goal of the control system designer is to improve the performance of the plant by controlling the plant's input.

A control system should:

Control systems, which are used to control the operation of other systems, are measured in terms of accuracy, speed of response, stability, and immunity from external noise.

1. Perform its function accurately
2. Respond quickly
3. Be stable
4. Respond only to valid inputs (noise immunity)

A system's accuracy determines how close the system's output will come to the desired output with a constant value input command. Quick response determines how closely the output of the system will track or follow a changing input command. A system's stability describes how a system behaves when a change, particularly a sudden change, is made by the input signal. Some unstable systems will oscillate wildly if uncontrolled. Others that are normally controlled may go out of control. Either case is undesirable and a good controller design will minimize the chance of unstable operation. A system should maintain its accuracy when noise or other disturbances are attempting to change the plant's output. These are invalid inputs and good design will eliminate them from system performance as much as possible. The more this invalid response is eliminated the more noise immunity the control system has.

Open Loop Control

In an open loop control system, the output of the system is not compared to the system's input, or control signal.

Accuracy, quick response, stability and noise immunity are all determined by the control system chosen for a particular plant. The basic control system block diagram is given in *Figure 2-11*. Here the command input is sent to a system block which performs some control operation on the input to generate an intermediate signal which drives the plant. This type of control is called open loop control because the output of the system is never compared with the command input to see if they match.

**Figure 2-11.
Open Loop Control
System Block Diagram**

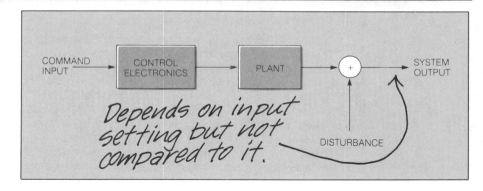

Closed Loop Proportional Control

In a closed loop control system, a feedback signal is used to compare a sampling of the output of the system to the input of the system. The resultant signal is used to attempt to reduce the error to zero.

A closed loop controller, shown in *Figure 2-12*, has all of the parts of an open loop controller plus an error amplifier which compares the system's output (normally measured by some sensor) with the input command. The error is determined by subtracting the output from the input. This error signal is then input to the control electronics where an appropriate plant control signal is determined based on the amount of error. Thus, the control signal is proportional to the error signal. This *feedback* arrangement enables the control electronics to monitor the effects of its control signal and respond more precisely to variations in the system's output.

Closed loop control improves the ability to satisfy all of the control system design goals. Accuracy, response time, stability and noise immunity can be improved because the control electronics can use the system's output to determine if it is satisfying the requirements of the input. Disturbances in the open loop system are transferred directly to the output, but in the closed loop system, the effect of the disturbance can be adjusted by changing the control gain. The control gain also affects stability, accuracy and quickness of response in a similar manner.

**Figure 2-12.
Closed Loop Control
System Block Diagram**

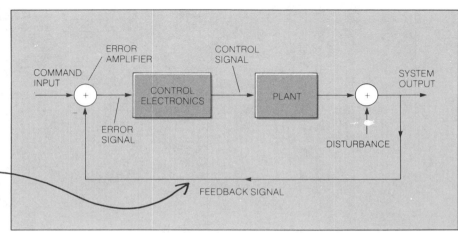

Closed Loop Limit Cycle Control

The control electronics in the previous example provided what is called proportional control because the control signal is proportional to the error signal. Other combinations of control electronics are possible and it is a challenge to the system designer to develop imaginative types of control electronics types to improve the performance of a given plant. Besides proportional control, another type of control which is used in automotive applications is called limit cycle control. Limit cycle control is a type of feedback control which monitors the system's output and responds only when the output goes beyond preset limits. Limit cycle controllers often are used to control plants with nonlinear or complicated transfer functions.

One example of a limit cycle controller is the temperature controlled oven depicted in *Figure 2-13*. The temperature of the oven is measured with a temperature probe and the corresponding electrical signal is fed back as an error signal. The temperature inside the oven is controlled by the length of time the heating coil is energized. The control electronics checks the error signal to determine if one of the following two conditions exist:

1. Oven temperature is below minimum setting of command input.
2. Oven temperature is above maximum setting of command input.

The control electronics responds to condition 1 by closing the relay contacts to energize the heating element. This causes the temperature in the oven to increase until the temperature rises above a maximum limit, condition 2. In

Limit cycle control responds only when the output goes beyond certain limits, thus it commonly cycles between upper and lower limits of control.

**Figure 2-13.
Limit Cycle Controller
to Control Oven
Temperature**

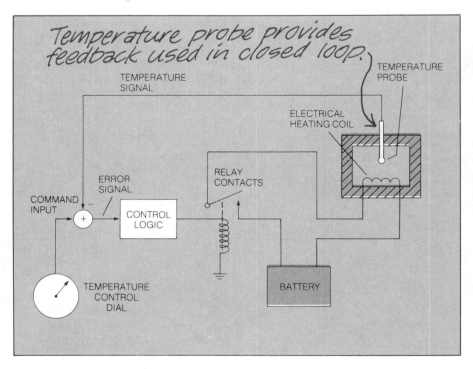

that case, the control electronics opens the relay contacts and the heat is turned off. The oven gradually cools because of heat loss until condition 1 again occurs and the cycle repeats. The oven temperature varies between the upper and lower limit and the variations can be graphed as a function of time as shown in *Figure 2-14*. The amplitude of the temperature variations, called the differential, can be decreased by reducing the difference between the maximum and minimum temperature limits. As the limits get closer together, the temperature cycles more rapidly (frequency increases) to hold the actual temperature much closer to the desired constant temperature. Thus, the limit cycle controller controls the system so as to maintain an average value close to the command input. This type of controller has gained popularity due to its simplicity, lower cost, and case of application. One of the most important automotive electronic control systems (i.e., fuel control) is a limit cycle controller (see Chapter 6).

Figure 2-14.,
Oven Temperature Graph

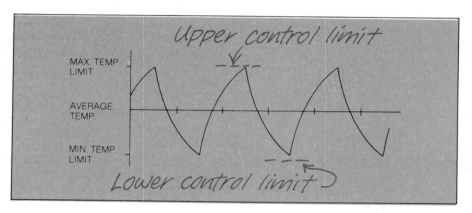

SUMMARY

The systems fundamentals and instrumentation and control basics covered in this chapter will be used in the remaining chapters so it is important to understand them. Modern automotive instrumentation and control systems are becoming more complex and sophisticated as new technologies are developed. Car buyers and the government are continually demanding greater precision and reliability from the various systems which measure and control an automobile's operation. The automobile industry is moving to meet this challenge of precision and reliability with electronic instrumentation and control systems.

Recently, automotive engineers have applied computer technology to solve some of the tougher problems. To prepare for the discussions of electronic and microcomputer automotive systems in later chapters, fundamentals of electronics will be presented in Chapter 3 and microcomputer fundamentals in Chapter 4.

Quiz for Chapter 2

1. Which of the following are examples of systems?
   a. clock
   b. electric dishwasher
   c. communication network
   d. society
   e. all of the above

2. Block diagrams are developed in what type of analysis?
   a. psychiatric
   b. quantitative
   c. qualitative

3. Specific system parameters are determined in what type of analysis?
   a. physical
   b. quantitative
   c. qualitative

4. What is a sensor used for?
   a. converts a non-electrical input to an electrical output
   b. converts an electrical input to a mechanical output
   c. reduces the effects of noise and other disturbances on the measured quantity
   d. all of the above

5. What does an actuator do?
   a. converts a non-electrical input to an electrical output
   b. converts an electrical input to an action
   c. reduces the effects of noise and other disturbances on the measured quantity
   d. all of the above

6. What does signal processing do?
   a. converts a mechanical input to an electrical output
   b. converts an electrical input to a mechanical output
   c. reduces the effects of noise and other disturbances on the measured quantity
   d. all of the above

7. An error amplifier is used to compare which of the following two signals?
   a. the output of a sensor and the input to a signal processor
   b. the output of a system and the command input of a system
   c. the output of a signal processor and the output of an actuator
   d. none of the above

8. A basic instrumentation system consists of which of the following components?
   a. sensor
   b. actuator
   c. signal processor
   d. all of the above

9. A control system may contain which of the following components?
   a. error amplifier
   b. control logic
   c. plant
   d. all of the above

10. Which of the following are examples of a plant?
   a. automotive drivetrain
   b. high temperature oven
   c. an airplane navigation system
   d. all of the above

Electronic Fundamentals

ABOUT THIS CHAPTER

As indicated in Chapter 1, the role of electronics in the automobile has changed dramatically from the original use of electric spark igniters to the present day use of electronic drive train controllers. Electronics is now used in place of some less efficient, less precise, less versatile, less convenient or more costly mechanical automotive subsystems.

This chapter is for the reader who has little knowledge of electronics. It is intended only to provide an overview of the subject so that discussions in later chapters about the operation and use of the automotive electronics control systems will be easier to understand. Readers that feel confident in the subject may wish to bypass this chapter or use it for a quick review.

The chapter is about electronic devices and circuits which are used in electronic automotive instrumentation and control systems. Topics include; semiconductor devices, analog circuits, digital circuits and the fundamentals of integrated circuits.

SEMICONDUCTOR DEVICES

Semiconductor devices are made from silicon or germanium which is purposely contaminated with impurities that change the conductivity of the material. Transistors are semiconductor devices that are used as active devices in electronic circuits.

Early electronic circuits used vacuum tubes as active devices (devices that provide a gain greater than one); today, solid-state transistors provide the circuit gain. A solid-state diode is similar to a transistor, but the diode does not provide gain. They are called solid-state because diodes and transistors are made from a solid material. The material in pure form is neither a good conductor nor a good insulator; therefore, it is called a semiconductor material, and transistors and diodes are called semiconductor devices.

Silicon and germanium are semiconductor materials commonly used to make diodes, transistors, and other semiconductor devices. Silicon is by far the most used material. The semiconductor material is made to conduct better by diffusing into it an impurity. Boron and phosphorous often are used as impurity source materials to alter the conductivity of silicon. When boron is used, the semiconductor material becomes a p-type semiconductor. When phosphorous is used, the semiconductor material becomes an n-type semiconductor.

Diodes and transistors are quite different from common linear components such as resistors, capacitors, inductors and transformers. Diodes and transistors are nonlinear devices. More care must be taken when analyzing their circuit performance.

Diodes

A diode is another kind of semiconductor device. It can be thought of as a one-way electrical street because current can flow through a diode in only one direction.

Diodes can be thought of as one-way resistors or current check valves because they allow current to flow through them in only one direction. In the forward (conducting) direction with a plus voltage on the p-type material (anode), diodes are a fairly good conductor with a forward resistance of about 5 to 10 ohms. This is called the forward biased condition. (The conventional current flow direction of positive to negative is used in this book.) In the reverse (nonconducting) direction with a plus voltage on the n-type material (cathode), diodes are good insulators with a resistance of a million ohms or more. This is called the reverse biased condition.

Figure 3-1a shows the schematic symbol for a diode and *Figure 3-1b* shows the actual and ideal V-I transfer characteristics. Notice on the ideal curve that the diode doesn't start conducting until the voltage across it exceeds V_d volts, then for small increases in voltage, the current increases very rapidly. For silicon diodes, V_d is about 0.7 volt. For germanium diodes, it is about 0.3 volt. Even for the actual curve, the change in current is quite steep for 0.1V changes in the voltage across the diode after Vd has been exceeded.

When designing or analyzing circuits, V_d is often ignored in relatively high voltage circuits where V_d is a very small percentage of the total voltage; however, in low voltage and low-level signal circuits, Vd may be a significant factor.

Rectifier Circuit

In rectifier circuits, diodes are used to block current flow during one-half of an ac cycle to convert ac to dc.

Figure 3-1c shows a very common diode circuit; consider it first without the dotted-in capacitor. The alternating current voltage source is a sine wave with a peak-to-peak amplitude of 100 volts (50 volt positive swing and 50 volt negative swing). Waveforms of the input voltage and output voltage plotted against time are shown in *Figure 3-1d* as the solid lines. Notice that the output never drops below 0 volts. The diode is reverse biased and blocks current flow when the input voltage is negative, but when the input voltage is positive, the diode is forward biased and permits current flow. Note that if the diode direction is reversed in the circuit, current flow will be permitted when the input voltage is negative, and blocked when the input voltage is positive.

The circuit of *Figure 3-1c* is called a half-wave rectifier because it effectively cuts the ac waveform in half. It is used to convert an alternating current (ac) voltage which goes above and below zero volts into a direct current (dc) voltage which stays either above zero volts or below zero volts, depending on which way the diode is installed. Rectifier circuits are commonly used to convert the ac voltage from a wall outlet into a dc voltage for use in battery chargers, radios and other appliances which require direct current.

**Figure 3-1.
Diode**

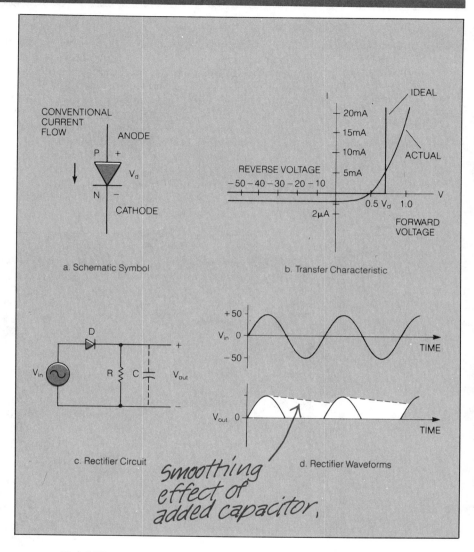

CONVENTIONAL
CURRENT
FLOW

ANODE

P +

V_d

N

CATHODE

a. Schematic Symbol

I

IDEAL

20mA

15mA

10mA

ACTUAL

REVERSE VOLTAGE

5mA

−50 −40 −30 −20 −10

V

0.5 V_d 1.0

2µA

FORWARD
VOLTAGE

b. Transfer Characteristic

D

+

V_{in} R C V_{out}

−

c. Rectifier Circuit

+50

V_{in} 0

−50

TIME

V_{out} 0

TIME

d. Rectifier Waveforms

smoothing effect of added capacitor,

The use of a capacitor to store charges and resist voltage changes smooths the rippling or pulsating output of a half-wave rectifier.

V_{in} of *Figure 3-1c* is ac; V_{out} is dc, but it is "bumpy" rather than smooth and steady. The bumpy output voltage of the half-wave rectifier can be smoothed by adding a capacitor as shown in dotted lines on *Figure 3-1c*. Since the capacitor stores charge and opposes voltage changes, it discharges (supplies current) to the load resistance (symbolized by R) when V_{in} is going negative from its peak voltage. The capacitor is recharged when V_{in} comes back to its positive peak and current is supplied to the load by the input voltage. The result is a V_{out} that is more nearly a smooth, steady dc voltage as shown by the dotted lines between the peaks of *Figure 3-1d*.

Transistors

Unlike diodes, transistors
are active elements that
can be used to strengthen,
or amplify, a signal. Transistors are three terminal
elements which act as
electrical valves.

Each of the two PN junc-
tions within a transistor
behaves much like the PN
junction in a diode.

Diodes and resistors are static circuit elements; that is, they don't
have gain or store energy. Transistors are active elements because they can
amplify or transform a signal level. Transistors are three terminal circuit
elements which act like current valves. There are two common bipolar types.
Figure 3-2a shows the schematic symbol for an NPN transistor and *Figure
3-2b* shows the schematic symbol for a PNP transistor. P represents p-type
semiconductor material and N represents n-type. The area where the p-type
and n-type materials join is called a PN junction, or just junction. Current
flows into the base and collector of an NPN transistor and out of the emitter.
The currents in a PNP transistor are exactly opposite to the NPN; that is,
current flows into the emitter and out of the base and collector. In fact, this is
the only difference between the PNP and NPN transistor. Their functions as
amplifiers and switches are the same.

**Figure 3-2.
Transistor Schematic
Symbols**

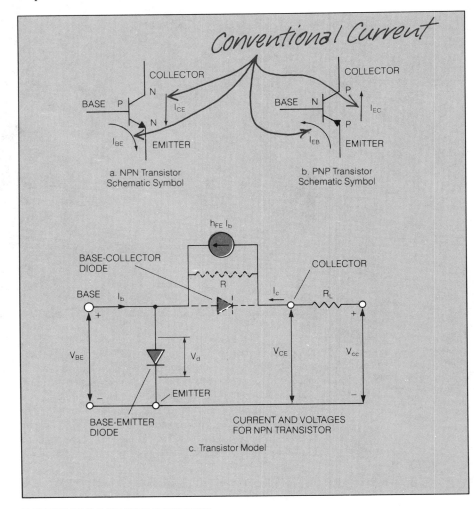

a. NPN Transistor Schematic Symbol

b. PNP Transistor Schematic Symbol

c. Transistor Model

During normal operation, current flows from the base to the emitter in an NPN transistor. The collector-base junction is reverse-biased, so that only a very small amount of current flows between the collector and the base when there is no base current flow.

The base-emitter junction of a transistor acts like a diode. Under normal operation for an NPN transistor, current flows forward into the base and out the emitter but does not flow in the reverse direction from emitter to base. The arrow on the emitter of the transistor schematic symbol indicates the forward direction of current flow. The collector-base junction also acts as a diode, but supply voltage is always applied to it in the reverse direction. This junction does have some reverse current flow, but it is so very small (1×10^{-6} to 1×10^{-12} amperes), that it is ignored except when operated under extreme conditions, particularly temperature extremes. In some automotive applications, the extreme temperatures may significantly affect transistor operation. For such applications, the circuit may include components that automatically compensate for changes in the transistor operation.

Under normal linear (analog) circuit operation, the collector-base junction is reverse biased as mentioned above; however, when used as a switch in the ON condition, the collector-base junction can become forward biased.

Transistor Model

To aid in circuit analysis, *Figure 3-2c* shows the diagram of a commonly used transistor model for an NPN transistor. The base-emitter diode is shown in solid lines in the circuit while the collector-base diode is shown dotted because generally it can be ignored.

The base-emitter diode does not conduct (there is no transistor base current) until the voltage across it exceeds V_d volts in the forward direction. If the transistor is a silicon transistor, $V_d = 0.7$ volt just like a silicon diode. The collector current, I_c, also is zero until V_{be} exceeds 0.7 volt. This is called the cutoff condition. It is the OFF condition when the transistor is used as a switch.

When V_{be} rises above 0.7 volt, the diode conducts and allows some base current, I_b, to flow. The collector current, I_c, is equal to the base current, I_b, times the transistor current gain, h_{FE}. h_{FE} can range from 10 to 200 depending on the transistor type. It is represented by a current generator in the collector circuit. This condition is called the active region because the transistor is on and amplifying. Also, it is called the linear region. The dotted resistance in parallel with the collector-base diode represents the leakage of the reversed biased junction and it is normally neglected, as discussed previously.

A transistor is saturated when a large increase in the base to emitter current results in only a small increase in the collector current.

A third condition called saturation exists under certain conditions of collector-emitter voltage and collector current. In the saturation condition, large increases in the transistor base current produce little increase in collector current. When saturated, the voltage drop across the collector-emitter is very small, usually less than 0.5 volt. This is the ON condition for a transistor switching circuit. This condition occurs in a switching circuit when the collector of the transistor is tied through a resistor, R_L, to a supply voltage, V_{cc} as shown in *Figure 3-2c*. Enough base current is supplied to the transistor to drive the transistor into the saturated condition where the output voltage (voltage drop from collector to emitter) is very small and the collector-base diode may become forward biased.

TRANSISTOR AMPLIFIERS

In a transistor amplifier, a small change in base current results in a corresponding, larger change in collector current.

Figure 3-3 shows a transistor amplifier. The ac voltage source, v_{in}, supplies a signal current to the base-emitter circuit. The transistor is biased to operate in the linear region at some steady state I_b and I_c. The voltage source, V_{cc}, supplies the steady state dc currents, I_b and I_c, and any signal i_c current change to the collector-emitter circuit. The small signal, v_{in}, varies the base current around the steady dc operating point. This small current change is i_b and it causes a corresponding but larger change in collector current, i_c, around the steady-state operating current I_c. The small signal current change causes an output voltage change v_{out} across the load resistor R_c. The small signal voltage gain of the circuit is as shown in *Figure 3-3*:

$$A = \frac{v_{out}}{v_{in}} = h_{fe}\frac{R_c}{R_b}$$

This is found by using the model and the equations, $v_{out} = i_c R_c$, $i_c = (h_{fe})i_b$, and $i_b = V_{in}/R_b$, where h_{fe} is the small signal current gain.

Circuits such as these are combined to make many types of amplifiers used in a variety of applications. Such circuits, especially when made in one package with integrated circuit technology (to be discussed later) are called linear circuits or analog circuits.

Transistor amplifiers are commonly used in analog circuits, including those in automotive systems.

Analog circuits are all around —in tape recorders, stereos, TV's, cassette players. They are used to simulate physical quantities and can be used in analog computers to model systems. They are used extensively in automotive systems. Analog circuits require a lot of time and care to design, so once an amplifier design works properly, the same circuit is copied and used in other systems when possible. The use of these standard building block circuits in complicated systems greatly reduces design time as well as reducing manufacturing and testing costs.

**Figure 3-3.
Transistor Amplifier
Circuit**

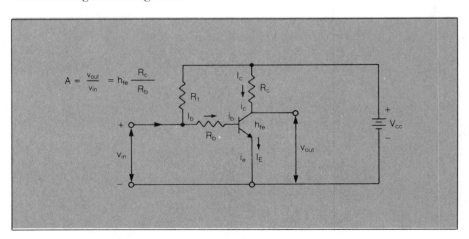

Operational Amplifiers

An op amp is an analog circuit that allows a building block approach to system design. Op amps offer high gain and other characteristics that can be tailored to match the requirements of a particular system.

The operational amplifier is such a building block that is popular because of its simplicity and the fact that it can be used in many applications. An operational amplifier (op amp) is a standard analog circuit that gives the circuit designer a package of gain whose characteristics can be varied easily to adjust it to the requirement of a particular system. Op amps typically have a very high voltage gain of 10,000 or more and typically have two inputs and one output as shown in *Figure 3-4a*. A signal applied to the inverting input (–) is amplified and inverted at the output. A signal applied to the noninverting input (+) is amplified, but is not inverted at the output.

Use of Feedback

The op amp is normally not operated at maximum gain, but feedback techniques are used to adjust the gain to the value desired as shown in *Figure 3-4b*. Some of the output is fedback to the input to oppose the input changes. The gain is adjusted by the ratio of the two resistors and is calculated by:

$$A_v = \frac{-R_f}{R_i} = \frac{v_{out}}{v_{in}}$$

Figure 3-4.
Operational Amplifiers

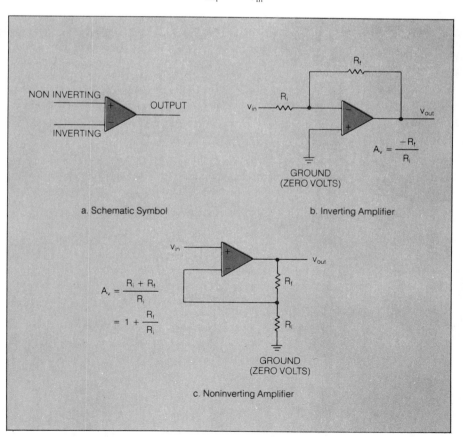

a. Schematic Symbol

$$A_v = \frac{-R_f}{R_i}$$

GROUND
(ZERO VOLTS)

b. Inverting Amplifier

$$A_v = \frac{R_i + R_f}{R_i}$$
$$= 1 + \frac{R_f}{R_i}$$

GROUND
(ZERO VOLTS)

c. Noninverting Amplifier

As indicated by applying the signal to the (−) terminal, the minus sign in the equation means signal inversion from input to output; that is, if the input goes positive, the output goes negative.

Since the op amp amplifies the voltage difference between its two inputs; it can be used as a differential amplifier as well as a single input amplifier.

A noninverting amplifier is also possible as shown in *Figure 3-4c*. The input signal is connected to the noninverting (+) terminal and the output is fed back through a series connection of resistors to the inverting (−) terminal. The gain, A_v, in this case is:

$$A_v = \frac{v_{out}}{v_{in}} = 1 + \frac{R_f}{R_i}$$

Besides adjusting gain, the negative feedback also can help to correct for the amplifier's non-linear operation and distortion.

Analog Computers

Analog computers, like those used to simulate the performance of automotive systems, are constructed with operational amplifiers.

The op amp is the basic building block for analog computers. Analog computers are used to simulate the behavior of other systems. The system's block diagram can easily be copied using the summers, multipliers and gain blocks of Chapter 2 made with op amps. Virtually any system that can be described in a block diagram using standard building blocks can be duplicated on an analog computer. If a control system designer is building an automotive speed controller and doesn't want to waste a lot of time and money testing prototypes on a real car, he can program the analog computer to simulate the car's speed electronically. By varying amplifier gains, frequency response, resistor, capacitor and inductor values, system parameters can be varied to study their effect on the system performance. Such system studies help to determine the parts and their characteristics needed for a system before any hardware is built.

The main problem with analog circuits and analog computers is that their performance changes with changes in temperature, supply voltage, signal levels, and noise level. Most of these problems are eliminated when digital circuits are used.

DIGITAL CIRCUITS

Circuits with outputs having either one of two possible predetermined states are called digital circuits.

Binary digital circuits are circuits whose output can be only one of two different states. Each state is indicated by a particular voltage or current level. An example of a simple binary digital system is a door open indicator. When your car door is opened, a light comes on. When it is closed, the light goes out. The system's output (the light from the bulb) is either on or off. The on state means the door is open, the off state means it is shut. (Binary means two valued. This will be discussed more later.)

Digital circuits also can use transistors. In a digital circuit, a transistor is in either one of two modes of operation: on, conducting at saturation; or, off, in the cutoff state.

In electronic digital systems, a transistor is used as a switch. Remember that the transistor has three operating regions; cutoff, active, and saturation. If only the saturation and cutoff regions are used, the transistor acts like a mechanical switch. When in saturation, it is ON and has very low resistance; when in cutoff, it is OFF and has very high resistance. The control input to the transistor switch must be capable of either saturating the transistor or turning it off without allowing operation in the active region. In the model of *Figure 3-2c*, the ON condition was indicated by a very low collector to emitter voltage and the OFF condition by a collector to emitter voltage equal to the supply voltage.

Binary Number System

The on-off states common in digital circuits can represent the two values (1 and 0) of the binary number system. The binary number system is called a base 2 system because it uses two digits to represent all numbers.

An important characteristic of a digital circuit is that the ON-OFF states can represent a 1 or 0 which are the two values used in the binary number system. To better understand the binary number system, let's compare the binary system with the decimal system in use everyday.

The binary number system uses only two digits, 0 or 1, and is called a base 2 system. The decimal system uses 10 digits, 0 through 9, and is called a base 10 system. In the decimal system, numbers are grouped from right to left with the first digit representing the ones place (10^0), the second digit the tens place (10^1), the third digit the hundreds place (10^2), and so on. Each place increases in value by a power of 10.

In the binary system, numbers are also grouped from right to left. The right most digit is in the ones place (2^0) and, because only the numbers 0 and 1 can be represented, the second digit must be the twos place (2^1), the third digit the fours place (2^2), the fourth digit the eights place (2^3) and so on. Each place increases in value by a power of 2. *Table 3-1* shows a comparison of place values. *Table 3-2* shows the binary equivalent for some decimal numbers. The binary number 0010 is read as "zero, zero, one, zero"; not "ten".

**Table 3-1.
Comparison of Place
Values**

—Also called digit position.

	DECIMAL-Base 10				BINARY-Base 2				
PLACE	4	3	2	1	5	4	3	2	1
VALUE	1000	100	10	1	16	8	4	2	1
POWER of BASE	3	2	1	0	4	3	2	1	0

Table 3-2.
Comparison of Numbers
in Different Bases

DECIMAL Base 10	BINARY Base 2
0	0000
1	0001
2	0010
3	0011
4	0100
5	0101
6	0110
7	0111
8	1000
9	1001
10	1010
11	1011
12	1100
13	1101
14	1110
15	1111
16	10000
255	11111111
256	100000000

To convert from binary to decimal, just multiply each binary digit by its place value and add the products. For instance, the decimal equivalent of the binary number 1010 is given by:

$$1010_2 = (1 \times 8) + (0 \times 4) + (1 \times 2) + (0 \times 1)$$
$$= 8 + 2$$
$$= 10_{10}$$

1010_2 means that the number is a base 2 or binary number. 10_{10} means the number is a base 10 or decimal number. Normal notation eliminates the subscripts 2 and 10.

Converting from decimal to binary can be accomplished by finding the largest number that is a power of 2 (divisor) which will divide into the decimal number (dividend) with a one as a quotient, putting a one in its place, and subtracting the number used to divide with (divisor) from the decimal number (dividend) to get a remainder. The operation is repeated by dividing with the next lower number that is a power of two until the binary ones place has been tested. Any time the dividend is less than the divisor, a zero is put in that place and the next power of 2 divisor is tried. For instance, to find the binary equivalent for the decimal number 73, the largest number that is a power of two and that will divide into 73 with a quotient of 1 is 64 (2^6):

$$(2^6) \qquad \frac{73}{64} = 1 \qquad 73 - 64 = 9$$

$$(2^5) \qquad \frac{9}{32} = 0$$

$$(2^4) \qquad \frac{9}{16} = 0$$

$$(2^3) \qquad \frac{9}{8} = 1 \qquad 9 - 8 = 1$$

$$(2^2) \qquad \frac{1}{4} = 0$$

$$(2^1) \qquad \frac{1}{2} = 0$$

$$(2^0) \qquad \frac{1}{1} = 1 \qquad 1 - 1 = 0$$

therefore; $\qquad\qquad 73 = 1001001$

LOGIC CIRCUITS (Combinational)

Digital computers can perform the binary digit (bit) manipulations very easily by using three basic logic circuits or gates. These gates are called the NOT gate, the AND gate and the OR gate.

NOT Gate

A NOT gate inverts its input signal level. If the input of a NOT gate is high, the output will be low, and vice versa.

The NOT gate is an inverter. If the input is a logic 1, the output is a logic 0. If the input is a logic 0, the output is a logic 1. It changes zeros to ones and ones to zeros. The transistor inverting amplifier of *Figure 3-3* performs this same function if operated from cutoff to saturation. A high base voltage (logic 1)[1] produces a low (logic 0) collector voltage and vice versa. *Figure 3-5a* shows the schematic symbol for a NOT gate. Next to the schematic symbol is what is called a "truth table". The truth table lists all the possible combinations of input A and output B for the circuit. The logic symbol is shown also. The logic symbol is read as "NOT A".

AND Gate

An AND gate requires all input signal levels to be high for the output signal to be high.

The AND gate is slightly more complicated. The AND gate has at least two inputs and one output. The one shown in *Figure 3-5b* has two inputs. The output is high (1) only when both inputs are high (1). If either or both inputs are low (0), the output is low (0). *Figure 3-5b* shows the truth table, schematic, and logic symbol for this gate. The two inputs are labeled A and B. Notice that there are four combinations of A and B, but only one results in a high output.

[1]Positive logic defines the most positive voltage as logic 1. Negative logic defines the most positive voltage as logic 0. Positive logic is used throughout this book.

**Figure 3-5.
Basic Logic Gates**

Provides output conditions for all combinations of inputs.

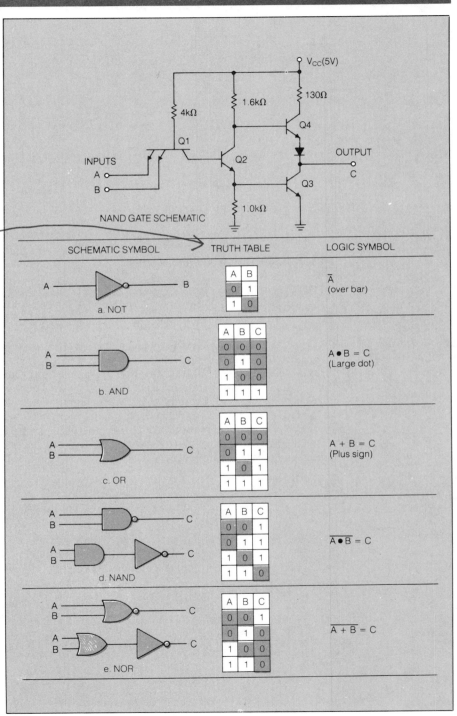

NAND GATE SCHEMATIC

SCHEMATIC SYMBOL	TRUTH TABLE	LOGIC SYMBOL

a. NOT

A	B
0	1
1	0

$\overline{A}$
(over bar)

b. AND

A	B	C
0	0	0
0	1	0
1	0	0
1	1	1

$A \bullet B = C$
(Large dot)

c. OR

A	B	C
0	0	0
0	1	1
1	0	1
1	1	1

$A + B = C$
(Plus sign)

d. NAND

A	B	C
0	0	1
0	1	1
1	0	1
1	1	0

$\overline{A \bullet B} = C$

e. NOR

A	B	C
0	0	1
0	1	0
1	0	0
1	1	0

$\overline{A + B} = C$

OR Gate

The output signal of an OR gate is high when any one of its input signal levels is high.

The OR gate, like the AND gate, has at least two inputs and one output. The one shown in *Figure 3-5c* has two inputs. The output is high (1) whenever one or both inputs are high (1). The output is low (0) only when both inputs are low (0). *Figure 3-5c* shows the schematic symbol, logic symbol and truth table for the OR gate.

NAND and NOR

NAND and NOR gates may be constructed by combining AND, OR, and NOT gates.

Other logic functions can be generated by combining these basic gates. An inverter can be placed after an AND gate to produce a NOT-AND gate. When the inverter is an integral part of the gate, the gate is called a NAND gate. The same can be done with an OR gate and the resultant gate is called a NOR gate. The truth table and schematic symbol for these gates are shown in *Figure 3-5d* and *Figure 3-5e*. Notice that the NOT function is indicated on the schematic symbol by a small circle at the output of each gate. The small circle is the schematic symbol for NOT and the overbar is the logic symbol for NOT. Notice also that the truth table outputs are the reverse of the AND and OR gate outputs. Where C was 1, it is now 0 and vice-versa. All of these gates are available in integrated circuit form with different quantities of gates in a package and a different number of inputs per gate. The schematic of an integrated circuit NAND gate is shown in *Figure 3-5*.

XOR and Adder Circuits

XOR gates, which output a high only when one or the other input is high, are commonly used to add binary numbers.

Another combination of gates performs the exclusive OR function, abbreviated as XOR. *Figure 3-6a* shows the schematic symbol, truth table and logic symbol. The output is high only when one input or the other is high, but not both. This gate very commonly is used for comparison of two binary numbers because if both inputs are the same, the output is zero. The equivalent combination of gates which performs this function also is shown in *Figure 3-6a*. The XOR gate is also available in an integral package so it is not necessary for the designer to interconnect separate gates to build the function.

These gates can be used to build digital circuits which perform all of the arithmetic functions of a calculator. *Table 3-3* shows the addition of two binary bits in all the combinations that can occur:

**Table 3-3.
Addition of Binary Bits**

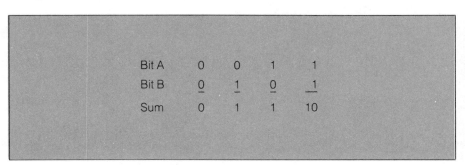

Bit A	0	0	1	1
Bit B	0	1	0	1
Sum	0	1	1	10

**Figure 3-6.
XOR and Adders**

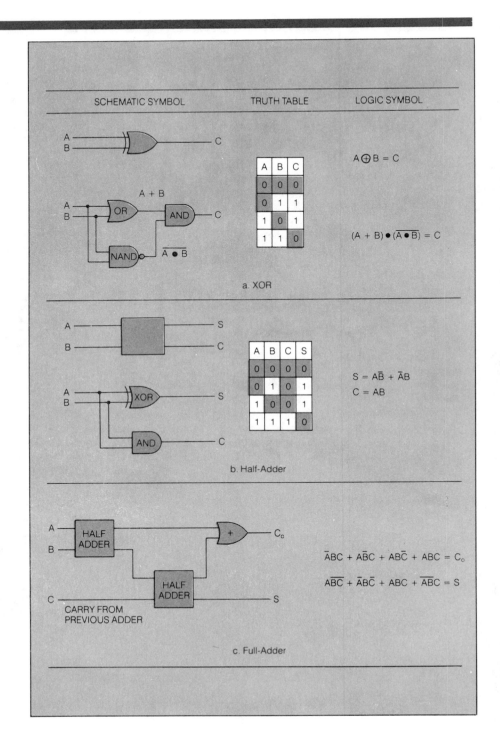

| SCHEMATIC SYMBOL | TRUTH TABLE | LOGIC SYMBOL |

a. XOR

A	B	C
0	0	0
0	1	1
1	0	1
1	1	0

$$A \oplus B = C$$

$$(A + B) \bullet (\overline{A \bullet B}) = C$$

b. Half-Adder

A	B	C	S
0	0	0	0
0	1	0	1
1	0	0	1
1	1	1	0

$$S = A\overline{B} + \overline{A}B$$
$$C = AB$$

c. Full-Adder

$$\overline{A}BC + A\overline{B}C + AB\overline{C} + ABC = C_o$$

$$A\overline{BC} + \overline{A}B\overline{C} + ABC + \overline{ABC} = S$$

CARRY FROM
PREVIOUS ADDER

Note that in the case of adding a 1 to a 1, the sum is zero, and a 1, called a carry, is placed in the next place value to be added with any bits in that place value. A digital circuit designed to perform the addition of two binary bits is shown in *Figure 3-6b*. It is called a half-adder. It produces the sum and any necessary carry as shown in the truth table.

A half-adder does not have an input to accept a carry from a previous place value. A circuit that does is a full-adder, shown in *Figure 3-6c*. A series of full-adder circuits can be combined to add binary numbers with as many digits as desired. A simple electronic calculator performs all arithmetic operations using full-adder circuits like these and a few additional logic circuits because subtraction is changed to addition, multiplication is repeated addition, and division is repeated subtraction.

LOGIC CIRCUITS WITH MEMORY (Sequential)

Sequential logic circuits have the ability to store, or remember, previous logic states. Sequential logic circuits are the basis of computer memories.

The logic circuits we have discussed so far have been simple interconnections of the three basic gates. The output of each system is determined only by the present inputs. These circuits are called combinational logic circuits. There is another type of logic circuit which has a memory of previous inputs or past logic states. This type of logic circuit is called sequential logic because the sequence of past input values and the logic state at the time determines the present output state. Because sequential logic circuits hold or store information even after inputs are removed, they are the basis of semiconductor computer memories.

R-S Flip Flop

A sequential logic circuit called an R-S flip flop can be set into either state. It will remain latched in that state until it is set to the opposite state by the presence of opposing logic signals on its two inputs.

A very simple memory circuit can be made by interconnecting two NAND gates, as in *Figure 3-7a*. A careful study of the circuit reveals that when S is high (1) and R is low (0), the output Q is set to a high and stays high regardless of whether S is high or low. The high state of S is said to be latched into the state of Q. The only way Q can be unlatched to go low is to let R go high and S go low. This resets the latch. This type of memory device is called a Reset-Set (R-S) flip-flop and is the basic building block of sequential logic circuits. The term "flip-flop" describes the action of the logic level changes at Q. Notice from the truth table that R and S must not be 1 at the same time. Under this condition, the two gates are bucking each other and the final state of the flip-flop output is uncertain.

J-K Flip Flop

A flip-flop where this uncertain state is solved is shown in *Figure 3-7b*. It is called a J-K flip-flop and can be obtained from an R-S flip-flop by adding additional logic gating as shown in the logic diagram. When both J and K inputs are 1, then the flip-flop changes to a state other than the one it was in.

**Figure 3-7.
Flip-Flops**

The basic memory building blocks.

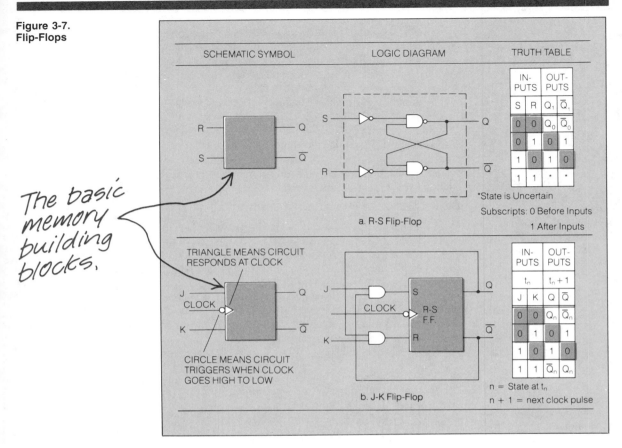

The J-K flip flop is a synchronized sequential logic circuit. It changes state only at a specific time; that time is determined by a timing pulse applied to a separate clock terminal.

The flip-flop shown in this case is a synchronized one. That means it changes state at a particular time determined by a timing pulse called the clock applied to the circuit at the terminal with a triangle. The little circle at the clock terminal means the circuit responds when the clock goes from a high level to a low level. If the circle is not present, the circuit responds when the clock goes from a low level to a high level.

Synchronous Counter

Figure 3-8 shows a four-stage synchronous counter. It is synchronous because all stages are triggered at the same time by the same clock pulse. It has 4 stages; therefore, it counts 2^4 or 16 clock pulses before it returns to a starting state. The timed waveforms appearing at each Q output also are shown. It is easy to see how such circuitry can be used for counting, for generating other timing pulses, and for determining timed sequences. One can easily visualize how such stages can be lined up to store the digits of a binary number. If the storage is temporary, then such a combination of stages is called a register. If storage is to be more permanent, it is called memory.

A succession of J-K flip flops can be combined to form a synchronous counter. Synchronous counters are the basis of digital clocks, and are used extensively in circuits that convert binary values to decimal numbers.

Digital clocks, as well as circuits that convert binary numbers to decimal numbers so they can be displayed and read by humans, are made up of many stages of such counting circuits. By using integrated circuit technology, all the counters, registers, and binary-to-decimal converters are produced, all at the same time, on a tiny piece of silicon semiconductor material. High performance circuits that use very low power, have high reliability, and are of a very small size are the results of integrated circuit technology.

Figure 3-8.
4-Stage Synchronous
Counter

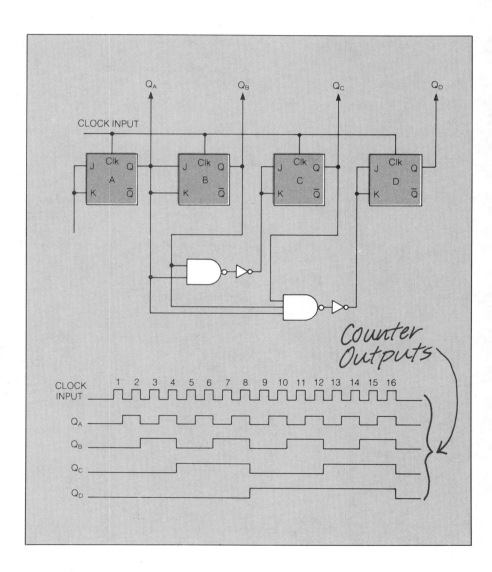

INTEGRATED CIRCUITS

Let's review what has been discussed on digital circuits. We know:

a. They operate with signals at discrete levels rather than with signals whose level varies continuously;

b. Voltage levels of low and high are commonly used to represent the binary numbers of 0 and 1;

c. Combinations of 1's and 0's can be used as codes to represent numbers, letters, symbols, conditions, etc.;

d. Circuits called gates (*Figures 3-5* and *3-6*) can be used in combination to make logical decisions;

e. Circuits called flip-flops (*Figure 3-7*) can be used to store 1's and 0's. They can be set or reset into particular binary sequences to produce or store digital information, or to count, or to produce timed digital signals;

f. Transistors are used in the ON and OFF condition in circuits to form gates and flip-flops.

ICs are ideal for digital circuits because the digital circuits consist of many interconnected identical gates.

Digital electronic systems send and receive signals made up of 1's and 0's in the form of codes. The digital codes represent the information that is moved through the digital systems by the digital circuits. Digital systems are made up of many *identical* logic gates and flip-flops interconnected to do the function required of the system. As a result, digital circuits are ideal to be made in integrated circuit form, because all components can be made at the same time on a small silicon area by using photolithography (photographic printing) and diffusion (modifying one material by combining it with another using high temperature) techniques. This is the heart of integrated circuit technology.

Overall Process

An integrated circuit is commonly referred to by its abbreviation, IC. The overall process for making IC's is shown in *Figure 3-9*. This process is a means of constructing and manufacturing circuits all at the same time in such a small space that the complete system made on a silicon chip fits on a fingertip, as shown in *Figure 3-9c*. The system has all of the resistors, transistors, diodes, capacitors and interconnections made into or on top of the silicon chip. A greatly magnified picture is shown in *Figure 3-9c*. This is the reason integrated circuits are called "fingertip electronics".

Figure 3-9.
Overall IC Process

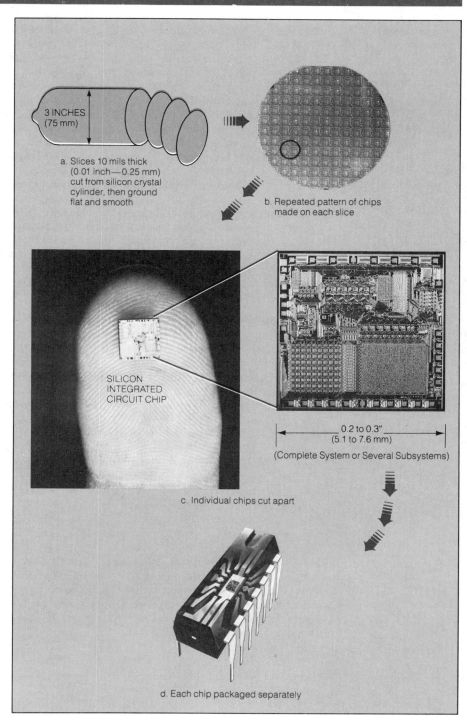

3 INCHES
(75 mm)

a. Slices 10 mils thick
   (0.01 inch—0.25 mm)
   cut from silicon crystal
   cylinder, then ground
   flat and smooth

b. Repeated pattern of chips
   made on each slice

SILICON
INTEGRATED
CIRCUIT CHIP

0.2 to 0.3″
(5.1 to 7.6 mm)

(Complete System or Several Subsystems)

c. Individual chips cut apart

d. Each chip packaged separately

Silicon crystals are the basic material for ICs. Individual slices are cut from the crystals, and the circuit patterns are etched into the surface of the silicon slice.

The overall process begins by growing a crystal of silicon which may be as large as two to six inches (5.08 to 15.2 cm) in diameter and four to twelve inches long. Round disks of nearly pure crystalline material (called a slice) are sliced from the cylinder crystal as shown in *Figure 3-9a*. Each slice, typically 10 mils [0.01 inch (0.25 mm)] thick, is ground and polished so that it looks like a mirror on one side. The combination of digital circuits forming a part of a system or a complete system in itself (such as the electronic engine controller that will be discussed in Chapter 6) is repeated in a regular pattern (*Figure 3-9b*) in a matrix of rows and columns to form hundreds of circuits on a single slice. The individual gates in the circuit are made in such a small space that hundreds of them can fit under the period at the end of this sentence. Some appreciation for this density can be obtained from the chip layout of *Figure 3-9c*. From 12,000 to 50,000 gates are put on chips of this size.

Once the circuits have been formed on the slice, the slice is cut apart with a diamond scribe to produce the individual chips, as shown in *Figure 3-9c*. Each chip is mounted on a metallized frame that supports it and provides the means for connecting the IC to other IC's or to other devices. The circuit connections on the chip are bonded to the metallized frame with 1 mil [0.001 inch (2.5 mm)] wires so that electrical connection can be made through the package pins as shown in *Figure 3-9d*. Plastic or ceramic seals the IC into the completed package.

Digital Evolution

The IC was invented by Jack Kilby of Texas Instruments in 1958. Texas Instruments and many other manufacturers have developed and manufactured a wide variety of many types of circuits. Both digital and linear (analog) circuits have been made. The first IC's contained only a few circuits, but as the technology improved, device sizes became smaller and higher levels of integration became possible. Complex circuits resulted from combinations of gates. More complex circuits resulted in more complex functions in a package. More complex functions were combined until subsystems, and finally complete systems, could be made in one IC.

Tables 3-4 and *3-5* show the progress of this digital evolution. *Table 3-4* shows how the density of gates on a chip has increased. In 1960, SSI (small scale integration) chips had from 10-12 gates. In late 1960, for MSI (medium scale integration), the density increased to 1,000. Through the 1970's, the density increased from LSI (large scale integration) to VLSI (very large scale integration) where the density increased to 50,000 gates or more. And progress will continue through the 1980's. The change in the number of gates from early 1960's through the 1970's has been an increase in density of 5,000 to 1; meaning, effectively, 5,000 times more digital information can be processed on the same size piece of silicon. All of this has been accomplished while still maintaining or improving system performance, decreasing size and weight, decreasing power requirements, providing wider operating temperature range, and increasing reliability.

**Table 3-4.
Summary of Digital
Evolution — Change in
No. of Gates**

Type	Time Period	No. of Gates	Change from SSI (ratio)
SSI	Early 1960	10-12	
MSI-LSI	Late 1960	100-1000	100:1
LSI-VLSI	Thru 1970	1000-50,000	5000:1

**Table 3-5.
Summary of Digital
Evolution — Change in
Chip Size**
(Source: Cannon, Luecke, Understanding Microprocessors, *Texas Instruments Incorporated, Copyright ©* 1979.

Type	Time Period	No. of Gates	AUG Chip Size		Chip Area (mils²)	Change from SSI (ratio)
SSI	Early 1960	10-12	(mils) 50x50	(mils) 1.3x1.3	2,500	
MSI-LSI	Late 1960	100-1000	150x150	3.8x3.8	22,500	9:1
LSI-VLSI	Thru 1970	1000-50,000	250x250	6.4x6.4	62,500	25:1

Circuit and Equipment Cost Impact

Increasing efficiencies and decreasing costs of ICs have resulted in widespread use of IC technology in automotive electronics.

By processing many slices at once, thousands of chips are made at the same time. It only costs a little more to make thousands of circuits than it does to make one. The individual chip cost is determined by how many are on a slice. The more chips that can be made from one slice, the lower the cost per chip. Therefore, silicon area is important. Look at *Tables 3-4* and *3-5* again. The chip size was reduced by 25 times from early 1960 to 1970, while at the same time, density was increased. Using both the number of gates per chip and the area reduction from these tables, they show that the area per gate of silicon used was reduced 200 times. As a result, cost per gate has been greatly reduced.

In early 1960, SSI IC's sold for $10, and had 10 gates; the cost per gate was $1. VLSI IC's with 50,000 gates may sell for $50 when they are first manufactured, to give a cost of 0.1¢ per gate; but as volume sales increase, the package cost decreases to about $10, for a gate cost of only 0.02¢ per gate—a reduction of 5,000 times! The automotive industry has the volume to take advantage of such tremendous cost reduction.

To demonstrate the impact on end equipment cost, look at *Table 3-6*. It shows what has happened to the cost of a medium-scale computer. Where in 1960, the hardware cost was around $30,000; in 1980, it was approximately $1,000. In 1985, it is projected to be only $100. That's a cost reduction of 300 times in this particular example.

This reduction in end equipment cost while providing much more functional capability, smaller size, lower power operation, and higher reliability are the reasons that automotive electronics are being manufactured using integrated circuit technology.

Table 3-6.
Cost of Medium-Scale
Computer
(Source: Cannon, Luecke,
Understanding
Microprocessors,
Texas Instruments
Incorporated, Copyright ©
1979.

Early 1960	$30,000
1970	$10,000
1977	$ 5,000
1980	$ 1,000
1985	$ 100

SUMMARY

In this chapter, the fundamentals and basic concepts of electrical circuits, and the components used in such circuits, have been discussed with emphasis on digital circuits and integrated circuit technology. In the next chapter, the same emphasis continues to show the expansion of digital circuits into computer systems. All of this to help you, the reader, better understand the functions performed by automotive electronics.

Quiz for Chapter 3

1. Forward conventional current flows in a diode circuit from:
   a. anode to cathode
   b. anode to anode
   c. cathode to anode
   d. cathode to cathode

2. Forward conventional current for a PNP transistor flows from:
   a. base to ground
   b. base to emitter
   c. emitter to base
   d. collector to base

3. The op amp is what type of circuit?
   a. digital
   b. analog
   c. logic gate
   d. none of the above

4. The AND gate is what type of circuit?
   a. digital
   b. analog
   c. amplifier
   d. inverter

5. Which conditions cause the output of an XOR gate to be high?
   a. both inputs are low
   b. both inputs are high
   c. either input is high but not both
   d. both inputs a zero

6. An R-S latch is what type of digital logic?
   a. combinational logic
   b. sequential logic
   c. Fortran
   d. J-K

7. Flip-flops are used in what type of logic systems?
   a. memories
   b. counters
   c. data registers
   d. all of the above

8. Integrated circuits are used in automotive electronic systems because they have:
   a. excellent functional performance
   b. high reliability
   c. small size
   d. low cost
   e. all of the above

9. VLSI integrated circuits may have circuit density equivalent to the functional capability of:
   a. 10-12 gates
   b. 1,000 gates
   c. 5,000 gates
   d. 50,000 gates

10. Integrated circuits' low cost is due to:
    a. processing many slices at a time
    b. high number of chips per slice
    c. processing one slice at a time
    d. no parts are the same
    e. a and b above

11. Which of the following materials is a semiconductor?
    a. iron
    b. lead
    c. silicon
    d. none of the above

12. Which of the following devices require the use of semiconductors?
    a. transistors
    b. diodes
    c. integrated circuits
    d. all of the above

13. In which of the following circuits is a diode used?
    a. filter
    b. rectifier
    c. resistor
    d. capacitor

14. What device is used to smooth
out the bumpy output of the
rectifier circuit?
 a. capacitor
 b. resistor
 c. diode
 d. transistor

15. In which type of circuits are
transistors used?
 a. amplifiers
 b. op-amps
 c. logic gates
 d. all of the above

16. Which of the following conditions
cause the output of an OR gate
to be low?
 a. both inputs are high
 b. only one input is high
 c. both inputs are low

17. Which of the following conditions
cause the outputs of an AND gate to
be high?
 a. both inputs are low
 b. both inputs are high
 c. one input is low
 d. at least one input is low

18. What decimal number does the
binary number 0110 represent?
 a. 4
 b. 3
 c. 110
 d. 6

19. What binary number does the
decimal number 10 represent?
 a. 0010
 b. 0101
 c. 1010
 d. 1000

20. The binary addition of 0110 and 0010
produces what binary sum?
 a. 1000
 b. 0111
 c. 1111
 d. 1010

Microcomputer Instrumentation and Control

ABOUT THIS CHAPTER

Ever since the first electronic computing machine was demonstrated in the 1950's, the computer has been having more and more effect on our daily lives. People use the computer to vote, bank, play games, research, learn, and find compatible mates. Computers have been programmed to control traffic flow on busy streets, route subways and trains, schedule arrivals and departures at airports, and many other similar activities. Space launches and missions are directed or controlled by computers. Space probes such as the ones to Saturn and Mars have computers on board which can be programmed by remote control to direct tests and control the path they travel. These uses of the complex combination of digital circuits called a computer have demonstrated its ability to perform under very demanding conditions; therefore, it is no wonder the automobile industry has enlisted its aid in their quest for improved fuel economy and emissions control.

This chapter is about microcomputers and how they are used in instrumentation and control systems. Topics include microcomputer fundamentals, microcomputer equipment, microcomputer inputs and outputs, computerized instrumentation, and computerized control systems.

MICROCOMPUTER FUNDAMENTALS

Digital Versus Analog Computer

In Chapter 3, we discussed analog computers that use continuous electrical signals to represent physical quantities so that the performance and functions of physical systems can be simulated. By simulating the system with an analog computer, the response of a physical system can be studied and predicted without building the system.

A digital computer also can be programmed to perform the same simulation. However, the inputs to the digital computer must be digital; that is, in discrete steps or at discrete levels, and the outputs will also be in discrete steps or levels. In the most common digital circuits, these discrete steps are usually one of two levels—called binary because of two levels—which are represented by the binary digits (bits) of one or zero ("1" or "0"). The analog computer is comprised of signal adders, multipliers and other analog circuitry that principally deal with continuously varying signals. The digital computer is comprised of digital circuitry of logic gates, binary adders, multipliers, data latches, memory circuits and other logic circuits. Digital computers were developed initially to overcome the problems of temperature drift and noise disturbance which affect analog computers.

Digital computers use binary signals, which are at one of two levels, to perform all operations. Digital circuits are affected less by heat and noise than the analog circuits used in analog computers.

Parts of a Computer

The parts of a digital computer are shown in *Figure 4-1*. The central processing unit (CPU) is the processor, and when made in an integrated circuit, it is called a microprocessor. It is where all of the arithmetic and logical decisions are made. This is the calculator part of the computer.

Memory holds the program and data. The computer can change the information in memory by writing new information into memory, or it can obtain information in memory by reading the information from memory. Each memory location has a unique address which the CPU uses to find the information it needs.

Information from people must be put into computer form so the computer can handle it, or the computer outputs must be outputted in a form so that humans can understand it. The input and output devices, called peripherals, do these conversions. Peripherals are devices such as paper tape readers and punches, card readers and punches, keyboard/CRT (cathode ray tube) terminals, magnetic tape units, magnetic disk units, and printers. The arrows on the interconnection lines indicate the flow of data.

A digital computer consists of a CPU to process information, a memory to store information, and input/output sections to communicate with the user.

**Figure 4-1.
Basic Computer Block Diagram**

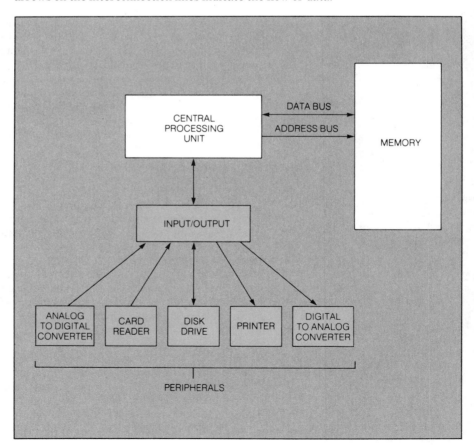

Microcomputer Versus Main Frame Computer

Microcomputers cost less and occupy less space than the mainframe computers commonly used by governments and large businesses. However, the microcomputers operate slower and are less accurate in mathematical operations.

With this general idea of what a computer is, let's see how a computer and a microcomputer differ. A microcomputer is just a small computer, typically thousands of times smaller than the large general purpose main frame computer used by banks and large corporations. The cost of a microcomputer is small compared to the cost of main frames and the computing power and speed is only a fraction of that of a main frame. A typical main frame computer costs from tens of thousands of dollars to millions of dollars and is capable of hundreds of thousands of arithmetic operations per second (additions, subtractions, multiplications and divisions). A microcomputer can be purchased for under $1,000.00 and can perform several thousand operations per second. More important for mathematical calculations than the speed of operation is the accuracy of the operation. Main frame computers use up to 64 bits to obtain high accuracy when doing arithmetic. The decimal equivalent for the largest number that can be represented using 64 bits is roughly 10 to the 19th power (1 followed by 19 zeros). A typical engine control microcomputer does arithmetic using only 8 bits. The largest decimal number that can be represented in 8 bits is 127 if one of the bits is used as a sign bit to indicate whether the number is positive or negative.

The computing power of microcomputers is far less than that of main frame computers, but they were never intended to be replacements for these "number crunching" machines. Microcomputers were designed to be general purpose replacements for tasks requiring only a limited computing capacity. As digital integrated circuit complexity increases, the capability of microcomputers will increase.

Programs

A program is a set of steps (instructions) in a logical order. The computer follows these steps to perform a given task.

A program is a set of instructions organized into a particular sequence to do a particular task. The first computers were little more than fancy calculators. They did only simple arithmetic and logical decisions. They were programmed (given instructions) by punching special codes into a paper tape which was read by the machine and interpreted as instructions.

A program containing thousands of instructions running on a machine such as this might require yards of paper tape. The computer would process the program by reading an instruction from the tape, perform the instruction, read another instruction from the tape, and so on until the end of the program. Reading paper tape is a slow process compared to the speed with which a computer can perform the requested functions. Also, if a program is to be run many times, the tape must be fed through each time. This is cumbersome and holds the possibility of the tape wearing and breaking.

Stored Programs

To minimize the use of paper tape, a method was invented to temporarily store programs inside the computer. The paper tape is read into a large electronic memory made out of thousands of data latches (flip-flops), one for each bit, which provide locations in which to store program instructions and data. Each instruction on the paper tape is converted to binary numbers with a definite number of bits and stored in a memory location. Each memory location has an address number associated with it like a post office box. In fact, you could think of the computer memory bank as a large bank of post office boxes. The computer reads the binary number (instruction or data) stored in each memory location by going to the address (box number) of the location it wants to read. When that location sees that it is being addressed, it responds by passing a *copy* of its information to the computer. (Note that the original information stays in the location when the memory is being read.) The electronic memory can be read much quicker than paper tape, so after the initial loading of the memory from the tape, the program can be run over and over without wasting the time required for reading paper tape.

Storing a computer program inside the computer's memory is what separates a real computer from a fancy calculator. The computer can use some of its memory for storing programs (instructions) and other memory for storing data. The program or data can be easily changed simply by loading in a different paper tape. The *stored program concept* is fundamental to all modern electronic computers.

We should also note that paper tape is only one method of loading programs and data and, in fact, isn't used much anymore. The more common ways are punched cards, keyboard terminals, magnetic tape, and magnetic disks.

WHAT CAN A MICROCOMPUTER DO?

A microcomputer can do the work of many different types of logic circuits. More importantly, the logic circuits a computer replaces are permanently wired to perform the function they were designed to do. If the design requirements are changed slightly, an entire printed circuit board or many boards may have to be redesigned to accommodate the change. With a microcomputer performing the logic functions, most changes can be made simply by reprogramming the computer. That is, the software (program) is changed rather than the hardware (logic circuits). This makes the microcomputer a very attractive building block in any digital system.

Microcomputers can also be used to replace analog circuitry. Special interface circuits can be used to enable a digital computer to input and output analog signals. We'll talk more about this a little later on and discuss how the conversion between analog and digital signals is made. The important point here is microcomputers present a person who wants to use digital systems with a very excellent alternative to hard wired (dedicated) logic and analog circuitry that is interconnected to satisfy a particular design. With this in mind, let's see how microcomputers work.

All modern electronic computers have the ability to store a program in internal memory. After a program is loaded into the computer's memory from a tape reader, disk drive, etc., the computer can use the program over and over to accomplish tasks.

Microcomputers offer flexibility because they can handle different tasks by using different programs. By comparison, ordinary logic circuits must be redesigned to handle different tasks.

HOW DOES A MICROCOMPUTER WORK?

A microcomputer stores information on a temporary basis within the CPU registers. Information is transferred between the CPU registers and the memory or input/output sections by means of one or more sets of multiple wires; each set is called a bus.

Recall the basic computer block diagram of *Figure 4-1*. The central processing unit (CPU) can request information from memory (or from an input or output device in most kinds of computer design) by calling it by its memory address. The address with all its bits is stored in the CPU as a binary number in a temporary data latch type memory called a register. The outputs of the register are sent at the same time over multiple wires to the computer memory and peripherals. As shown in *Figure 4-2*, the group of wires which carry the address is called the address bus. (The word bus refers to one or more wires that is a common path to/from many places.) The address register used in most microcomputers holds 16 bits which enable the CPU to access 65,536 memory locations. Each memory location usually contains 8 bits of data. A group of 8 bits is called a byte and a group of 4 bits is sometimes called a nibble.

Figure 4-2.
Buses and Registers

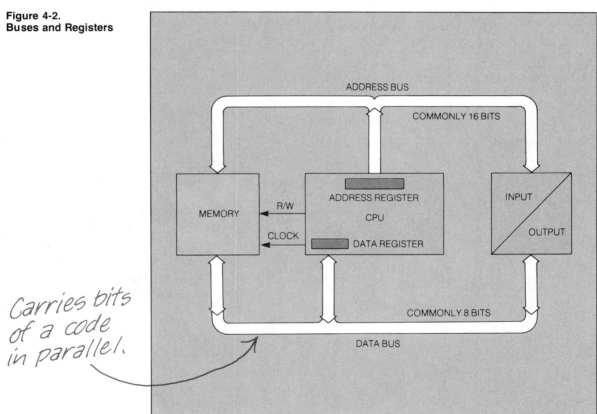

Carries bits of a code in parallel.

Information is sent to or received from memory locations and input/output devices via the bidirectional data bus.

Data is sent to the CPU over a data bus (*Figure 4-2*). The data bus is slightly different from the address bus in that the CPU uses it to *read* information from memory or peripherals and to *write* information to memory or peripherals. Signals on the address bus originate only at the CPU and are sent to devices attached to the bus. Signals on the data bus can either be inputs to the CPU or outputs from the CPU. The information on the data bus is sent or received at the CPU by the data register. In other words, the data bus is a two-way "street" while the address bus is a one-way "street". Another difference is that the data bus in many microcomputers is only 8 bits wide but, as already mentioned, the address bus is 16 bits wide.

The CPU always controls the direction of data flow on the bus because, although it is bi-directional, data can move in only one direction at a time. The CPU provides a special read/write control signal (*Figure 4-3*) which tells the memory in which direction the data is flowing. For example, when the read/write (R/W) line is high, the CPU wants to read information from a memory location. When the R/W line is low, the CPU wants to write information into a memory location.

**Figure 4-3.
Timing Diagram for
Memory Read**

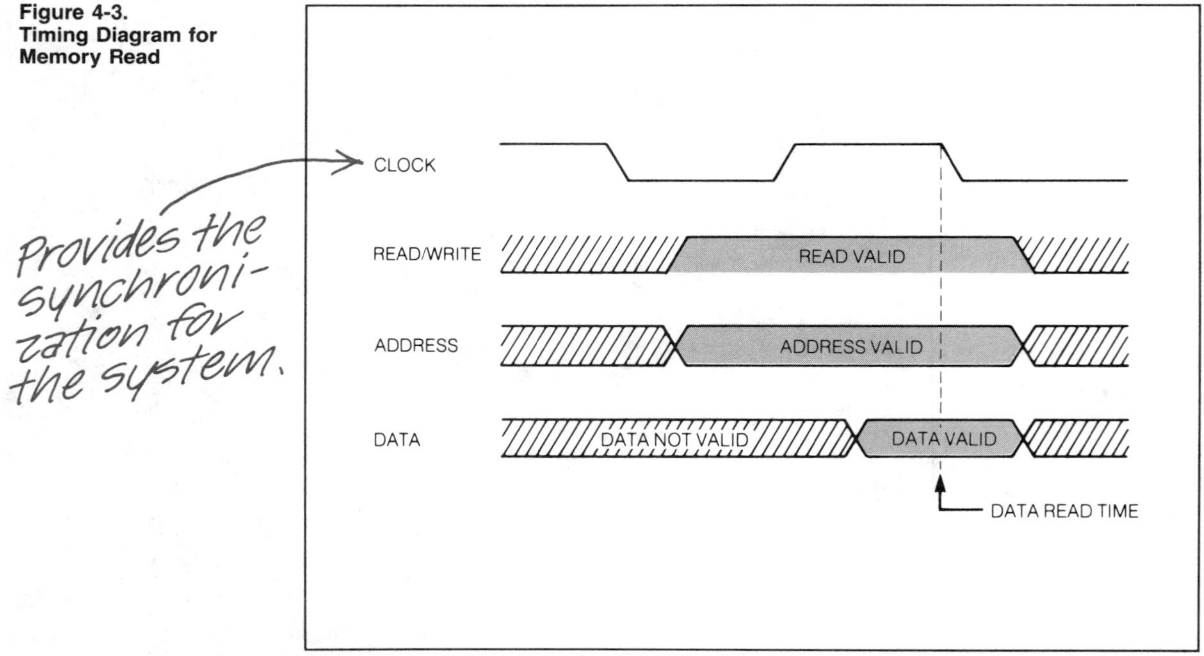

Provides the synchronization for the system.

Memory Read

During a memory read operation, the CPU changes the state of the read/write line and puts the appropriate address on the address bus. This causes the contents of the addressed memory location to be placed on the data bus, where it can be read by the CPU.

Let's take a detailed look at what happens on the microcomputer busses during a memory read operation. The timing diagram for a memory read operation is shown in *Figure 4-3*. Suppose the computer has been given the instruction to read data from memory location number 10. To perform the read operation, the CPU raises the R/W line to the high level to tell the memory to prepare for a read operation. Almost simultaneously, the address for location 10 is placed on the address bus (address valid in *Figure 4-3*). The number 10 in binary (0000 0000 0000 1010) is sent to the memory on the address bus. Memory location number 10 recognizes that its address has been called and quickly puts a copy of its contents (instruction or data) on the data bus.

Timing

Microcomputers use a timing signal, called a clock, to determine when data should be written to or read from memory.

A certain amount of time is required for the memory's address decoder to figure out (decode) which memory location is called for by the address and for the selected memory location to transfer its information to the data bus. To allow time for these things to happen, the processor waits awhile before it looks at the data bus for the information requested. Then at the proper time, the CPU opens the logic gating circuitry between the data bus and the CPU data register so that the information on the bus from memory location 10 is latched into the CPU. During the memory read operation, the memory has temporary control of the data bus. Control must be returned to the CPU, but not before the processor has read in the data. The CPU provides a timing control signal called the clock which tells the memory when it can take and release control of the data bus.

Refer again to *Figure 4-3*. Notice that the read cycle is terminated when the clock goes from high to low during the time that the read signal is valid. This is the signal the CPU uses to tell the memory that it has read the data and the data bus can be released. The timing for a memory write operation is very similar with the main difference being the R/W line is low instead of high.

The bus timing signals that have been discussed are very important to the reliable operation of the computer. However, they are under machine control and built into the design of the machine. Therefore, as long as the machine performs the read and write operations correctly, the programmer can completely ignore the logical details of the bus timing signals and concentrate on the logic of the program.

Addressing Peripherals

Microcomputers that use a design called "memory mapped" I/O send data to the peripherals in the same manner as data is sent to the memory.

Up to now, we have made a point of distinguishing between memory locations and peripherals. The reason for this is that they perform different functions. Memory is a data storage device while peripherals are input/output devices. Many microcomputers address memory and peripherals the same because they use a design called memory-mapped I/O (input/output). With this design, peripherals, such as data terminals, look just like memory to the CPU so that sending data to a peripheral is as simple as writing data to a memory location. In systems where this type microcomputer has replaced some digital logic, the digital inputs enter the computer through a designated memory slot. If outputs are required, they exit the computer through a designated memory slot.

CPU REGISTERS

The programmer (the person that writes the sequences of instructions for a particular task) uses a different model of the microprocessor used in the system than the hardware designer. The programmer uses a programming model. This model shows the programmer which registers in the CPU are available for program use and what function the registers perform. *Figure 4-4* shows a programming model for a typical 8-bit microcomputer. The model shows that the computer has two 8-bit registers and three 16-bit registers. The 16-bit registers will be discussed later on, but let's examine the 8-bit registers now.

Accumulator Register

The accumulator register, also called the A-register, is used for arithmetic and logical operations.

One of the 8-bit registers is an accumulator. An *accumulator* is a general purpose register which is used for arithmetic and logical operations. The accumulator can be loaded with data from a memory location or its data can be stored to a memory location. The number held in the accumulator can be added to, subtracted from, or compared with another number from memory. *The accumulator is the basic work register of a computer.* It is commonly called the A register.

**Figure 4-4.
Registers Available in a
Typical Microcomputer**

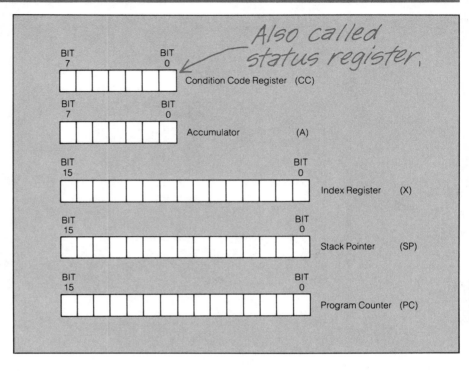

The condition code register (also called the status register) indicates certain conditions that take place during various accumulator operations.

Condition Code Register

The other 8-bit register, the condition code (CC) register (also called status register), indicates or flags certain conditions which occur during accumulator operations. Rules are established in the design of the microprocessor so that a 1 or 0 in the bit position of the condition code register represents specific conditions that have happened in the last operation of the accumulator. The bit positions and rules are shown in *Figure 4-5a*. One bit of the CC register indicates that the A register is all zeros. Another bit, the carry bit, indicates that the last operation performed on the accumulator caused a carry to occur. The carry bit acts like the ninth bit of the accumulator. Notice what happens when we add one to 255 in binary.

Decimal	Binary
255	11111111
+ 1	+ 1
256	100000000

The eight bits in the accumulator are all zeros, but the carry bit being set to a 1 (high) indicates that the result is actually not zero, but 256. Such a condition can be checked by examining the CC register carry bit for a 1.

Figure 4-5.
Use of the CC Register

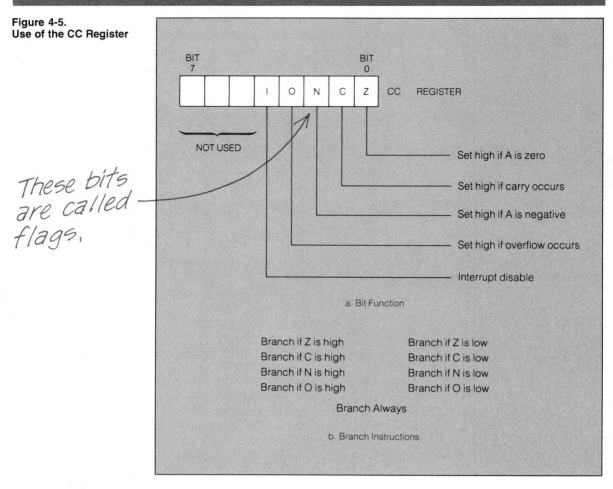

These bits are called flags,

a. Bit Function

Branch if Z is high   Branch if Z is low
Branch if C is high   Branch if C is low
Branch if N is high   Branch if N is low
Branch if O is high   Branch if O is low

Branch Always

b. Branch Instructions

For example, the condition code register can indicate if the value contained in the accumulator is negative. This gives the CPU the ability to represent a much wider range of numbers.

The condition code register also provides a flag which, when set to a 1, indicates that the number in the accumulator is negative. Most microcomputers use a binary format called 2's complement notation for doing arithmetic. In 2's complement notation, the left-most bit indicates the sign of the number. Since one of the 8 bits is used for the sign, 7 bits (or 15 if 16 bits are used) remain to represent the magnitude of the number. The largest positive number that can be represented in 2's complement with 8 bits is +127 (or +32,767 for 16 bits). The most negative number is −128 (−32,768). Since the accumulator is only 8 bits wide, it can handle only 1 byte at a time. However, by combining bytes and operating on them one after another in time sequence (as is done for 16-bit arithmetic), the computer can handle very large numbers or can obtain increased accuracy in calculations. Handling bits or bytes one after another in time sequence is called serial operation.

Branching

Instructions that direct
the microcomputer to
other parts of the program
are called branches.
Branches may be con-
ditional or unconditional.

The condition code register provides programmers with status indicators (the flags) which enable them to monitor what happens to the data as the program executes the instructions. The microcomputer has special instructions which allow it to go to a different part of the program. Bits of the CC register are labeled in *Figure 4-5a*. Typical branch type instructions are shown in *Figure 4-5b*. These program branches are either conditional or unconditional.

Eight of the nine branch instructions listed in *Figure 4-5b* are conditional branches. That is to say, the branch is taken only if certain conditions are met. These conditions are indicated by the CC register bit as shown. The branch-always instruction is the only unconditional branch. Such a branch is used to branch around the next instruction to a later instruction or to return to an earlier instruction. Another type instruction that takes the computer out of its normal program sequence is indicated for the I bit of the CC register. It is associated with an interrupt. An interrupt is a request, usually from an input or output (I/O) peripheral that the CPU stop what it's doing and accept or take care of (service) the special request. We'll have more about interrupts later in this chapter.

HOW DOES THE COMPUTER READ INSTRUCTIONS?

To understand how the computer performs a branch, we must first discuss how the computer reads program instructions from memory. Recall that the program instructions are stored in memory as binary numbers. The instructions are stored sequentially (step by step) starting at a certain binary address and ending at some higher address. The computer uses a register called the program counter (*Figure 4-4*) to keep track of where it is in the program.

Initialization

The program counter is
used by the computer to
keep track of where it is in
a particular program.

To start the computer, a small start-up (boot) program which is permanently stored in the computer sets all CPU registers with the correct values and clears all information in the computer memory to zeros before the program is loaded. This is called *initializing* the system. Then, the program to be run is loaded into memory and the address of the first program instruction is loaded into the program counter. The first instruction is read from the memory location whose address is contained in the program counter register; that is, the 16 bits in the program counter are used as the address for a memory read operation. Each instruction is read from memory in sequence and sent on the data bus into the instruction register where it is decoded. The instruction register is another temporary storage register inside the CPU. It is connected to the data bus when the information on the bus is an instruction.

Operation Code

The actual instructions in the program are in the form of numeric codes called operation codes (op codes).

Numeric codes in the instructions that represent the actual operation to be performed by the CPU are called operation codes (or op-codes for short). The block diagram of *Figure 4-6* showing part of the CPU hardware organization should help clarify the flow. The instruction register has a part which contains the numeric codes that represent the op-codes. A decoder determines the operation to be executed from the op-codes and a data router controls the flow of data inside the CPU as a result of the op-code.

**Figure 4-6.
CPU Organization**

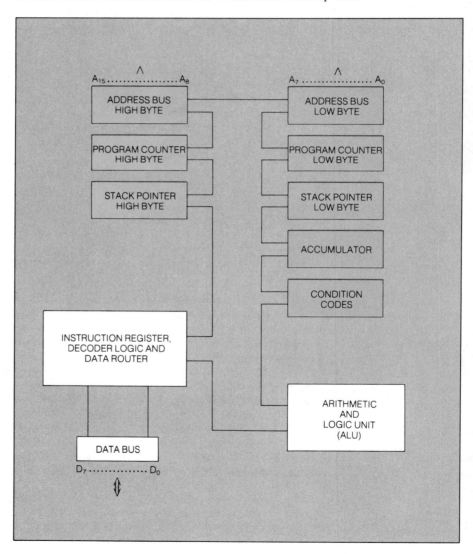

Instructions often are contained in more than one byte. In such cases, the first byte contains the op code, and succeeding bytes contain the address.

One important function of the op-code decoder is to determine how many bytes must be read to execute each instruction. Many instructions require two or three bytes. *Figure 4-7* shows the arrangement of the bytes in an instruction. The first byte contains the op-code. The second byte contains address information, usually the low or least significant byte. If there is a third byte, it usually contains the high or most significant byte of the address.

**Figure 4-7.
Instruction Byte
Arrangement**

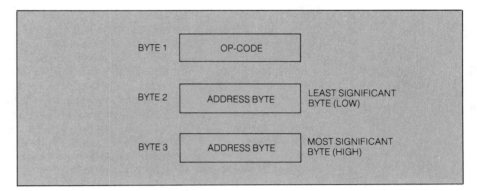

Program Counter

Each successive read of a memory location causes the program counter to be incremented to the address of the next byte.

The program counter is used by the CPU to address memory locations which contain instructions. Every time an op-code is fetched (read) from memory, the program counter is incremented (advanced by one) so that it points to the next byte following the op-code. If the operation code requires another byte, the program counter supplies the address, the second byte is fetched from memory and the program counter is incremented. Each time the CPU performs a fetch operation, the program counter is incremented; thus, the program counter always points to the next byte in the program. Therefore, after all bytes required for one complete instruction have been read, the program counter points to the beginning of the next instruction to be executed.

Branch Instruction

All of the branch instructions require two bytes. The first byte holds the operation code and the second byte holds the location to which the processor is to branch. When the instruction decoder decodes the first byte and finds that it is a branch instruction, it knows that a second byte must be read and that the second byte contains address information.

Now, if the address information associated with a branch instruction is only 8 bits long and totally contained in the second byte, then it can not be the actual branch address. In this case, the code contained in the second byte is actually a 2's complement number which the CPU adds to the lower byte of the program counter to determine the actual new address. This number in the second byte of the branch instruction is called an *address offset* or just offset.

A positive branch offset address results in a branch to a higher memory location, while a negative branch offset address results in a branch to a lower memory location.

Recall that in 2's complement notation, the 8-bit number can be either positive or negative; therefore, the branch address offset can be positive or negative. A positive branch offset causes a branch forward to a higher memory location. A negative branch offset causes a branch to a lower memory location. Since 8 bits are used, the largest forward branch is 127 and the largest backward branch is 128 memory locations.

Offset Example

Let's use a simplified example to clarify this. Suppose the program counter is at address 5,122 and the instruction at this location is a branch instruction. The instruction to which the branch is to be made is located at memory address 5,218. Since the second byte of the branch instruction is only 8 bits wide, the actual address 5,218 cannot be contained. Therefore, the difference or offset (96) between the current program counter value (5,122) and the desired new address (5,218) is contained in the second byte of the branch instruction. The offset value (96) is added to the address in the program counter (5,122) to obtain the new address (5,218), which is then placed on the address bus. The binary computation of the final address from the program counter value and second byte of the branch instruction is shown in *Figure 4-8.*

**Figure 4-8.
Binary Computation of
Branch Address**

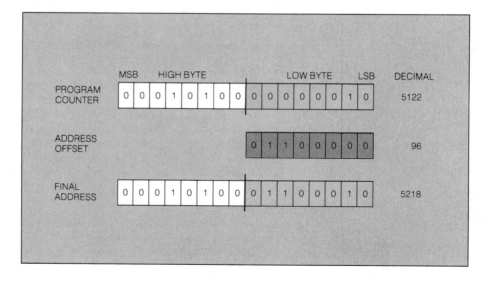

Jump Instruction

Eight-bit branch operations are limited to an offset range of + 127 or -128 memory locations. Thus, program branches to locations further away must use jump instructions. These 3-byte instructions contain the entire memory address.

As mentioned, the branch instructions have a range of + 127 or − 128. If the branch needs to be farther, a jump instruction must be used. The jump instruction is a 3-byte instruction. The first byte is the jump op-code and the next two bytes are the actual jump address. The CPU loads the jump address directly into the program counter and the program counter effectively gets restarted at the new jump location. The CPU continues to fetch and execute instructions in exactly the same way it did before the jump was made.

The jump instruction causes the CPU to jump out of one section of the program into another. The CPU cannot automatically return to the first section because no record was kept of the previous location. In other words, the jump instruction doesn't leave a trail for the CPU to follow back to the place from which it jumped. However, another instruction, the jump-to-subroutine, does leave a trail.

Jump-to-Subroutine Instruction

Subroutines are short programs used to perform specific tasks, particularly those tasks that must be performed several times within the same program.

A subroutine is a short program which is used by the main program to perform a specific function. It is located in sequential memory locations separated from the main program sequence. If the main program requires some function such as addition several times at widely separated places within the program, the programmer can write one subroutine to perform the addition, then have the main program jump to the memory locations containing the subroutine each time it is needed. This saves having to rewrite the addition program over and over again. To perform the addition, the programmer simply includes instructions in the main program which first loads the numbers to be added into the data memory locations used by the subroutine and then jumps to the subroutine.

The second and third bytes of a jump-to-subroutine instruction provide the address of the subroutine to be jumped to.

Refer to *Figure 4-9* to follow through the sequence. It begins with the program counter pointing to address location 100 where it gets the jump-to-subroutine instruction (step 1). When the instruction decoder encounters a jump-to-subroutine instruction (step 2), it knows that the next two bytes must also be read to obtain the jump address (step 2a). Therefore, the program counter is incremented once for each byte (steps 3 and 4) and the jump address is loaded into the address register. The program counter is then incremented once more so that it points to the op-code byte of the next instruction (step 5).

Saving the Program Counter

The contents of the program counter are saved by storing them in a special memory location before the jump address is loaded into the program counter. This program counter address is saved so the CPU knows where to return to in the main program when the subroutine is finished. This is the trail back that was mentioned before.

**Figure 4-9.
Jump-to-Subroutine**

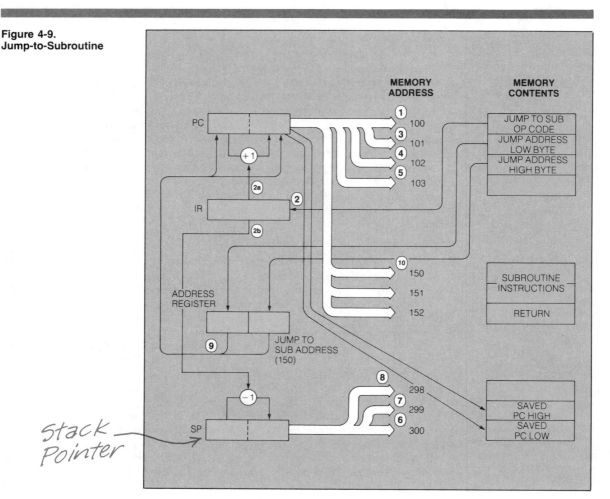

For a jump-to-subroutine, the contents of the program counter (after being incremented) are stored in two memory locations pointed to by the stack pointer. After storing it, the stack pointer value is decreased by one to prepare it for the next store.

Now refer back to *Figure 4-4*. There is a register in *Figure 4-4* called the stack pointer (SP). The address of the special memory location used to store the program counter is kept in this 16-bit stack pointer register. When a jump-to-subroutine op-code is encountered, the CPU uses the number code contained in the stack pointer as a memory address to store the program counter to memory (step 2b). The program counter is a two-byte register, so it must be stored in two memory locations. The current stack pointer is used as an address to store the lower byte of the program counter to memory (step 6). Then the stack pointer is decremented (decreased by 1) and the high byte of the program counter is stored in the next lower memory location (step 7). The stack pointer is then decremented again to point to the next unused byte in the stack to prepare for storing the program counter again when required (step 8).

When the subroutine is completed, a return instruction retrieves the saved program counter value from the stack pointer and loads it into the program counter. Execution of the main program then resumes from the point at which the jump occurred.

The special memory locations pointed to by the stack pointer are called a stack because, if you consider memory locations as being slots stacked one atop the other, the stack pointer stacks data like plates on a shelf.

As mentioned above, after the program counter has been incremented and saved, the jump address is loaded into the program counter (step 9). The jump to subroutine is made, and the CPU starts running the subroutine (step 10). The only thing that distinguishes the subroutine from another part of the program is the way in which it ends. When a subroutine has run to completion, it must allow the CPU to return to the point in the main program from which the jump occurred. In this way, the main program can continue without missing a step. The return-from-subroutine instruction is used to accomplish this. It is decoded by the instruction register and increments the stack pointer as shown in *Figure 4-10*, step 1. It uses the stack pointer to address the stack memory to retrieve the old program counter value from the stack (steps 2 and 4). The old program counter value is loaded into the program counter register (steps 3 and 5) and execution resumes in the main program (step 6). The return-from-subroutine instruction works like the jump-to-subroutine instruction except in reverse.

**Figure 4-10.
Return-from-Subroutine**

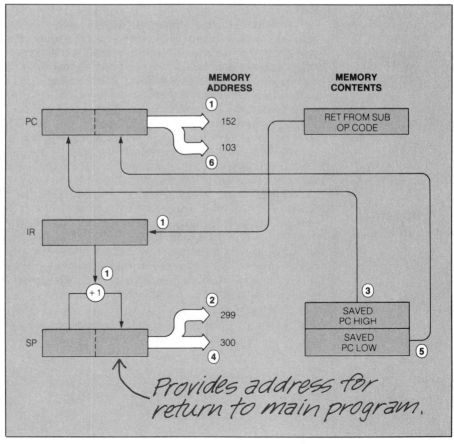

EXAMPLE USE OF A MICROCOMPUTER

Let's look at an example of how a microcomputer might be used to replace some digital logic and along the way learn about some more microcomputer instructions.

Microcomputers can be used in place of discrete logic circuits such as AND gates.

The digital logic to be replaced is a simple AND gate circuit. Now, no one would use a microcomputer only to replace an AND gate because an AND gate costs only a fraction of what a microcomputer costs. However, if the system already has a microcomputer in it, the cost of the AND gate could be eliminated by performing the logical AND function in the computer rather than with the gate. This is a perfectly legitimate application for a microcomputer and is something that microcomputers do very well.

Suppose there are two signals that must be ANDed together to produce a third. One of the input signals comes from a pressure switch under the driver's seat of an automobile to indicate whether someone is sitting in it or not. This signal will be called A and it will be at logic high when someone is sitting in the seat. Signal B will be at logic high when the driver's seat belt is fastened. The output of the AND gate is signal C. It will be logic high when someone is sitting in the driver's seat AND has the seatbelt fastened.

Buffer

Buffers provide temporary storage for peripheral inputs and let the microcomputer treat peripherals, such as sensors, as if they were memory locations.

In order to use a microcomputer to replace the AND gate, a means must be provided for the computer to detect the status of each signal. Recall that the computer knows only what is stored in its memory. The microcomputer used has memory-mapped I/O where peripherals are treated exactly like memory locations. The task is to provide a peripheral which will allow the computer to look at the switch signals as if they were bits in a memory location. This can be done easily by using a device called a buffer shown in *Figure 4-11*. To the microcomputer, a buffer looks just like an 8-bit memory slot at a selected memory location. The 8 bits in the memory slot correspond to 8 digital signal inputs to the buffer. Each digital input controls the state of a single bit in the memory slot. The digital inputs are gated into the buffer under control of the CPU. The microcomputer can detect the state of the digital inputs by examining the bits in the buffer any time after the inputs are gated into the buffer.

In this application, signal A will be assigned to the right-most bit (bit 0) and signal B to the next bit (bit 1). It doesn't matter that the other 6 bits are left unconnected. The computer will gate in and read the state of those lines, but the program will be written to purposely ignore them. With the logic signals interfaced to the microcomputer, we can begin writing the program which will perform the required logic function.

Figure 4-11.
Buffer

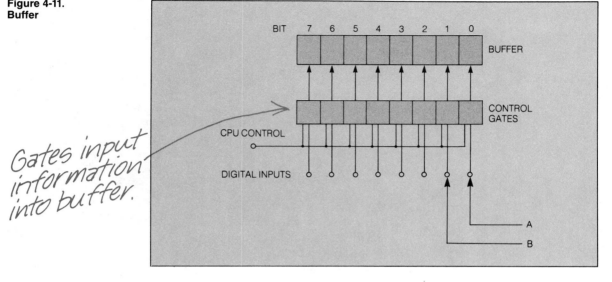

Gates input information into buffer.

Assembly Language

Microcomputer instruc-
tions are written in assembly language, a type of
shorthand that uses ini-
tials or shortened words to
represent microcomputer
instructions.

Microcomputer instructions are written in a special type of
abbreviated language called *assembly language*. Some assembly language
instructions such as branch, jump, jump-to-subroutine and return-from-
subroutine have already been discussed. Others will be discussed as they are
needed in our example program. Assembly language instructions have the
form of initials or shortened words which represent microcomputer functions.
These abbreviations are only for the convenience of the programmer because
the program that the microcomputer eventually runs must be in the form of
binary numbers. When each instruction is converted to the binary code that
the microcomputer recognizes, it is called a machine language program.

Once the program has
been written in assembly
language, a special kind of
program, called an as-
sembler, converts the as-
sembly-language program
into the binary code
recognized by the
microcomputer.

Assembly language instructions are called mnemonics (first m is
silent). The assembly language mnemonic for the jump instruction is JMP.
Table 4-1 shows the mnemonics for typical microcomputer instructions along
with a longer detailed description of the operation called by the instruction.
When writing a microcomputer program, it is easier and faster to use the short
mnemonic rather than the long function name. Assembly language just makes
things easier for the computer programmer because the mnemonics are easier
for a person to remember and write than the binary numbers the computer has
to use. However, the computer program must eventually be converted to the
binary codes that the microcomputer recognizes as instructions. A special
program, called an assembler, can be run on the computer to convert the
mnemonics to the binary codes they represent. This enables the programmer
to write the program using words that have meaning to the programmer and
produce machine codes that have meaning to the computer. Without going into
any more detail as to how assembler programs work, let's look at some of the
instructions that will be needed in the AND gate application program.

Mnemonic	Operand	Comment
JMP	(Address)	Jump to new program location
JSR	(Address)	Jump to a subroutine
BRA	(Offset)	Branch using the offset
BEQ	(Offset)	Branch if accumulator is zero
BNE	(Offset)	Branch if accumulator is non-zero
BCC	(Offset)	Branch if carry bit is zero
BCS	(Offset)	Branch if carry bit is non-zero
BPL	(Offset)	Branch if minus bit is zero
BMI	(Offset)	Branch if minus bit is non-zero
RTS		Return from a subroutine

a. Program Transfer Instructions

Mnemonic	Operand	Comment
LDA	(Address)	Load accumulator from memory
STA	(Address)	Store accumulator to memory
LDA	# (Constant)	Load accumulator with constant
LDS	# (Constant)	Load stack pointer with constant
STS	(Address)	Store stack pointer to memory

b. Data Transfer Instructions

Mnemonic	Operand	Comment
COM		Complement accumulator (NOT)
AND	(Address)	AND accumulator with memory
OR	(Address)	OR accumulator with memory
ADD	(Address)	ADD accumulator with memory
SUB	(Address)	SUBtract accumulator with memory
AND	# (Constant)	AND accumulator with constant
OR	# (Constant)	OR accumulator with constant
SLL		Shift accumulator left logical
SRL		Shift accumulator right logical
ROL		Rotate accumulator left
ROR		Rotate accumulator right

c. Arithmetic and Logical Operations

Logic Functions

A microcomputer can AND the contents of its accumulator with a memory location to perform the logical AND function.

Microprocessors are capable of performing all of the basic logic functions such as AND, OR, NOT and combinations of these. For instance, the NOT operation affects the accumulator by changing all ones to zeros and zeros to ones. Other logic functions are performed using the contents of the accumulator and some memory location. All eight bits of the accumulator are affected, and all are changed at the same time. As shown in *Figure 4-12*, the AND operation requires two inputs. One input is the contents of the accumulator and the other input is the contents of a memory location; thus, the eight accumulator bits are ANDed with the eight memory bits. The AND operation is performed on a bit-by-bit basis. For instance, bit 0 of the accumulator (the rightmost bit) is ANDed with bit 0 of the memory location, bit 1 with bit 1, bit 2 with bit 2, and so on. In other words, the AND operation is performed as if eight AND gates were connected with one input to a bit in the accumulator and the other to a bit (in the same bit position) in the memory location. The resulting AND outputs are stored back into the accumulator in the corresponding bit positions. The OR logical function is performed in exactly the same way as the AND except a 1 would be produced at the output if Signal A OR Signal B is a 1, or both are a 1.

Figure 4-12.
Microcomputer Logical
AND Function

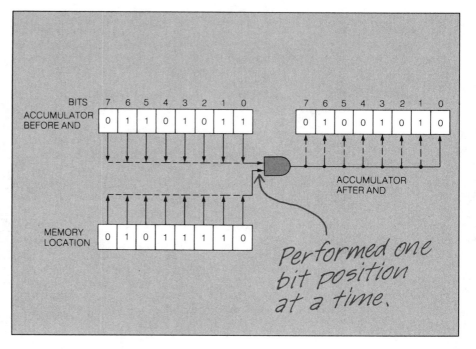

Shift

Using a logical operation known as SHIFT, a microcomputer can shift all the bits present in the accumulator to the left or right.

Instead of the AND gate inputs being switched to each bit position as shown in *Figure 4-12*, the microcomputer uses a special type of sequential logic operation, the shift, to move the bits to the AND gate inputs. A shift operation causes every bit in the accumulator to be shifted one bit position. The shift can be either right or left. It can be what is called a logical shift or it can be a circulating type shift. *Figure 4-13* shows the four types and their effect on the accumulator. In a left shift, bit 7 (the left-most bit) is shifted into the carry bit of the condition code register. Bit 6 is shifted into bit 7, and so on until each bit has been shifted once to the left. Bit 0 (the right-most bit) can be replaced either by the carry bit or by a zero, depending on the type of shift performed. There are two categories of shift operations shown providing a total of four types of shifts.

**Figure 4-13.
Types of Shift
Operations**

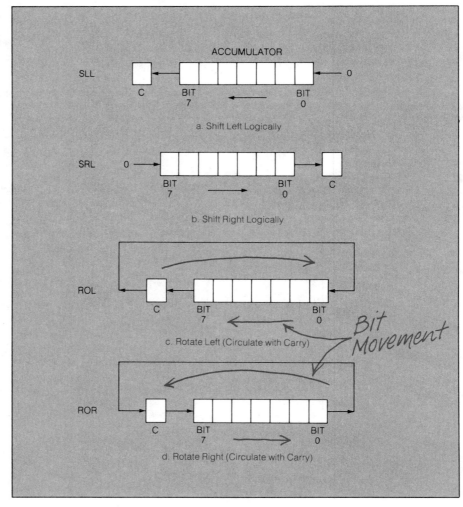

a. Shift Left Logically

b. Shift Right Logically

c. Rotate Left (Circulate with Carry)

d. Rotate Right (Circulate with Carry)

Programming the AND Function

It is the task of the programmer to choose instructions and organize them in such a way that the computer performs the desired functions. To program the AND function, one of the instructions will be the AND for "AND accumulator with memory" as shown in *Table 4-1c*. Since the AND affects the accumulator and memory, values must be put into the accumulator to be ANDed. This requires the load accumulator instruction, LDA.

The assembly language program of *Figure 4-14a* performs the required AND function. The programmer must first know which memory location the digital buffer interface (*Figure 4-11*) occupies. This location is identified and the programmer writes instructions in the assembler program that the buffer memory location will be referred to by the label or name SEAT. SEAT is easier for the programmer to remember and write than the address of the buffer.

An assembly-language program to cause the microcomputer to perform the logical AND function would use the LDA instruction to load the value to be ANDed into the accumulator.

Figure 4-14.
Assembly Language
Subroutines

Program Label	Mnemonic	Operand
1 CHECK	LDA	SEAT
2	AND	#00000001 B
3	SLL	
4	AND	SEAT
5	RTS	

a. Subroutine CHECK

Program Label	Mnemonic	Operand
1 WAIT	JRS	CHECK
2	BEQ	WAIT
3	RTS	

b. Subroutine WAIT

The use of a mask allows the microcomputer to separately examine two or more digital signals (bits) occupying the same 8-bit byte.

The operation of the program is as follows. The accumulator is loaded with the contents of the memory location SEAT. Note in *Figure 4-11* that the two digital logic input signals, A and B, have been gated into bits 0 and 1 respectively of the buffer which occupies the memory location labeled SEAT. Bit 0 is high when someone is sitting in the driver's seat. Bit 1 is high when the driver's seat belt is fastened. Only these two bits are to be ANDed together, the other six are to be ignored. But there is a problem because both bits are in the same 8-bit byte and there is no single instruction to AND bits in the same byte. However, the two bits can be effectively separated by using a mask.

Masking

During a mask operation, the accumulator contents is ANDed with the mask value which has a zero in each bit location except for the bit(s) to be saved. The saved, or masked, bit(s) comes through unchanged, but all others are set to zero.

Masking is a means to allow only selected bits to be effective in a desired operation. Since the buffer contents have been loaded into the accumulator, only bits 0 and 1 have meaning and these two bits are the only ones of importance that are to be kept in the accumulator. To do this, the accumulator is ANDed with a constant that has a zero in every bit location except the one that is to be saved. The binary constant in line 2 of *Figure 4-14a* (00000001) is chosen to select bit 0 and set all others to zero as the AND instruction is executed. This ANDing procedure is called *masking* because we have placed a window or mask over the accumulator which allows only bit 0 to come through unchanged. If bit 0 was a logic 1, it is still a logic one after masking. If bit 0 was a logic 0, it is still a logic 0. All other bits in the accumulator are set to zero by the masking operation. Therefore, the accumulator now contains the correct bit information about bit 0.

Shift and AND

During the final part of the AND operation, the Shift Left Logically instruction is used to align the bits of signal A with the correct bits in signal B so the logical AND can be accomplished.

However, the accumulator is still not ready to perform the final AND operation. Remember that SEAT contains the contents of the buffer and the condition of Signal A and Signal B. The contents of the accumulator must be ANDed with SEAT so that signals A and B are ANDed together. A copy of signal A is held in the accumulator in bit 0, but it is in the wrong bit position to be ANDed with signal B in SEAT in the bit 1 position. Therefore, signal A must be shifted into the bit 1 position. To do this the shift left logical instruction is used (*Figure 4-13a*). With signal A in bit 1 of the accumulator and signal B in bit 1 of SEAT, the AND operation can be performed on the two bits. If both A and B are high, then the AND operation will leave bit 1 of the accumulator high (1). If either is low, bit 1 of the accumulator will be low (0).

Subroutines Usefulness

This program has been written as a subroutine named CHECK so it can be used at many different places in a larger program. For instance, if the computer is controlling the speed of the automobile, it might want to be able to detect whether a driver is properly fastened in the seat before it sets the speed at 85 kilometers per hour.

Since the driver's seat information is very important, the main program needs to wait until the driver is ready before allowing anything else to happen. A program such as shown in *Figure 4-14b* can be used to do this. The main program calls the subroutine WAIT which in turn immediately calls the subroutine CHECK. CHECK returns to WAIT with the condition codes set as they were after the last AND instruction. The Z bit (*Figure 4-5a*) is set if A and B are not both high (the accumulator is zero). The BEQ instruction (*Table 4-1*) in line 2 of WAIT branches back because the accumulator is zero and causes the computer to re-execute the JSR instruction in line 1 of WAIT. This effectively holds the computer in a loop rechecking signals A and B until the accumulator has a non-zero value (A and B are high).

Timing Error

The time required for the microcomputer to sample sensor inputs and perform its instructions must be taken into account during program design; otherwise, timing errors may result.

A flaw in the subroutine CHECK could cause it to incorrectly perform the AND function. Notice that the logic state of A and B is sampled at different times. Signal A is first read in and masked off, then signal A and B are ANDed together. There is a possibility that during the interval between the time A is read and the time A and B are ANDed, the state of A could change. A computer is fast, but it still takes a certain amount of time for the microprocessor to execute the program instructions. For the driver's seat application, the signals have a long time between change so the time lag is not critical. However, in systems where the timing of signals is very tight, the program would have to be rewritten to remove the lag. Even after correcting such a lag, there may be applications where variables change more rapidly than the sampling time. Special compromises must be made in such cases or a new technique found to solve the problem.

This AND gate problem has revealed a number of important features of microcomputers. It shows how computers interface with other digital systems and how they sample and manipulate those signals. Let's now turn our attention to more detail on the components that make up the microcomputer system.

MICROCOMPUTER HARDWARE

The microcomputer system electronic components are known as computer hardware. (The programs the computer runs are called software.) The basic microcomputer parts were described as: the CPU, memory, and I/O (input and output peripherals).

CPU

The central processing unit is a microprocessor. It is an integrated circuit similar to the one shown in *Figure 4-15a*. It contains thousands of transistors and diodes on a chip of silicon small enough to fit on the tip of a finger. The chip is housed in a rectangular, flat package similar to the one shown in *Figure 4-15b*. It is less than 3 inches long and has two rows of small pins that make connections to external circuits. The CPU gets program instructions from a memory device.

Memory-ROM

Permanent memory, called ROM, maintains its contents even when power is turned off. ROMs are used for CPU instructions requiring permanent storage.

There are several types of memory devices available and each has its own special features. Systems, such as those found in the automobile, which must permanently store their program use a type of permanent memory called *Read Only Memory (ROM)*. This type of memory can be programmed only one time and the program is stored permanently, even when the microcomputer power is turned off. The programs stored in ROM are sometimes called *firmware* rather than software since they are unchangeable. This type of memory enables the microcomputer to immediately begin running its program as soon as it is turned on.

**Figure 4-15.
IC Chip Microprocessor**

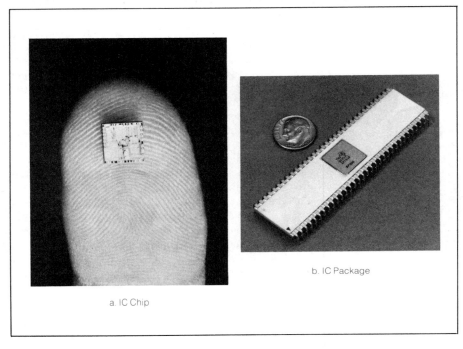

b. IC Package

a. IC Chip

Memory-RAM

A form of memory that can be written to and changed, as well as read from, is commonly referred to as RAM.

Another type of memory that can be written to as well as read from is required for the program stack, data storage, and program variables. This type of memory is called *Random Access Memory (RAM)*. This is really not a good name to distinguish this type of memory from ROM because ROMs are also random access type memories. Random access means the memory locations can be accessed in any order. They don't have to be accessed in any particular sequence. A better name for the data storage memory would be Read/Write Memory (RWM). However, the term RAM is commonly used to indicate a read/write memory, so that's what we'll use. A typical microcomputer contains both ROM and RAM type memory.

I/O-Parallel Interface

Microcomputers require interface devices which enable them to communicate with other systems. The digital buffer interface used in the driver's seat application discussed earlier is one such device. The digital buffer interface is an example of a parallel interface because the eight buffer lines are all sampled at one time or in parallel. The parallel buffer interface in the driver's seat application is an input or readable interface. Output or writeable interfaces allow the microcomputer to affect external logic systems. An output buffer must be implemented using a data latch so that the binary output is retained after the microcomputer has finished writing data into it. This permits the CPU to go on to other tasks while the external system reads and uses the output data. This is different from the parallel input where the states could change between samples.

Digital-to-Analog Converter

A DAC converts binary signals from the micro-computer to analog voltages that are proportional to the number encoded in the input signals.

The parallel input and output interfaces are used to examine and control external digital signals. The microcomputer can also be used to examine and control analog signals through the use of special interfaces. The microcomputer can produce an analog voltage by using a digital-to-analog converter (DAC). A DAC accepts inputs of a certain number of binary bits and outputs a voltage level which is proportional to the input number. DACs come in many different versions with different numbers of input bits and output ranges. The most common microcomputer DACs have 8-bit inputs and a 0-5 volt output range.

A simple 8-bit digital-to-analog converter is shown in *Figure 4-16*. This type of DAC uses a parallel input interface and two operational amplifiers. The 8 bits are written into the parallel interface and stored in data latches. The output of each latch is a digital signal which is zero volts if the bit is low and 5 volts if the bit is high. The first op-amp is a summing amplifier and has a gain of $-R_f/R_i$. The second op-amp has a gain of -1; thus, it is only an inverter. The effect of the two amplifiers is to scale each bit of the parallel interface by a specially chosen factor and add the resultant voltages together. For instance, if only bit 0 is high and all the others are low;

$$V_{out} = 5 \left[1 \left(\frac{1}{256} \right) + 0 \left(\frac{1}{128} \right) + 0 \left(\frac{1}{64} \right) + \ldots + 0 \left(\frac{1}{2} \right) \right]$$

$$V_{out} = \left(\frac{5}{256} \right)$$

$$V_{out} = 0.0195V$$

If only bits 0 and 7 are high;

$$V_{out} = 5 \left[1 \left(\frac{1}{256} \right) + 0 \left(\frac{1}{128} \right) + 0 \left(\frac{1}{64} \right) \ldots + 1 \left(\frac{1}{2} \right) \right]$$

$$V_{out} = \left(\frac{645}{256} \right)$$

$$V_{out} = 2.5195V$$

**Figure 4-16.
Digital-to-Analog
Converter Circuit Block
Diagram**

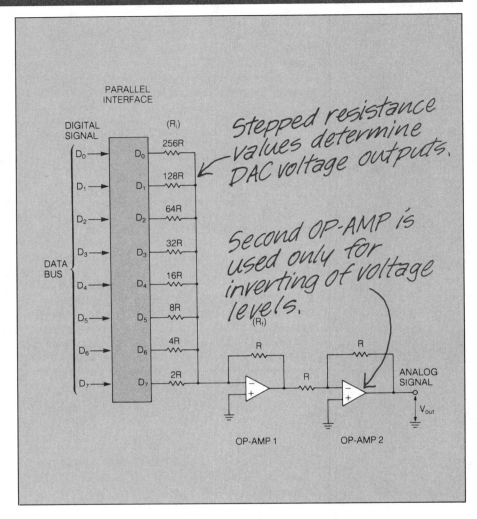

Stepped resistance values determine DAC voltage outputs.

Second OP-AMP is used only for inverting of voltage levels.

Strictly speaking, the DAC voltage output is still a digital signal because it can have only certain voltage levels in discrete steps. This causes the analog output voltage to have a staircase appearance as the binary number at the input is increased one bit at a time from minimum value to maximum value as shown in *Figure 4-17*. This DAC can have any one of 256 different voltage levels and for many applications, this is a close enough approximation to an analog signal.

Figure 4-17.
Staircase Output Voltage
of the DAC in Figure 4-16

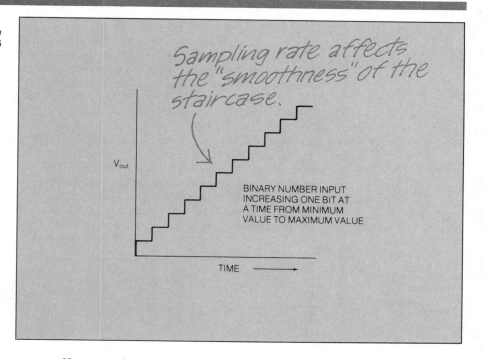

The accuracy of the DAC's representation of the digital input signal varies with circuit design, and with the rate at which the input signal is sampled.

However, there can be a problem with how fast the output can follow the input. The DAC output voltage can change only when the computer writes a new number into the DAC data latches. This is called the sampling rate of the input digital signal. The sampling rate must be often enough to ensure an accurate representation of the changes in the digital signal. The analog output of the digital-to-analog converter can take only a specific number of different values and can change only at specific times determined by the sampling rate. The output of the converter will always have small discrete step changes (resolution) and the designer must decide how small the steps must be to produce the desired shape and smoothness in the analog signal so that it is a reasonable duplication of the variations in the digital levels. In some systems, too low a sampling rate or not enough resolution may limit the usefulness of the DAC in controlling analog systems.

Analog-to-Digital Converter

Since a microcomputer can produce an analog voltage output from a binary number input (using a DAC), is the inverse true? It certainly is. Microcomputers can measure analog voltages by using a special interface component called an analog-to-digital converter (ADC). Analog-to-digital converters convert an analog voltage input into a digital number output that the microcomputer can read.

Figure 4-18 shows one way of making an ADC by using a DAC and a voltage comparator. The input to the DAC is a binary number which is generated by the microcomputer starting at the minimum value and increasing toward a maximum value just as in *Figure 4-17*. This binary number is generated at the parallel output of the microcomputer. The output of the DAC, V_{out}, is one input to the comparator. The other input is the input voltage, V_{in}, that the ADC is measuring. When the V_{out} voltage of the DAC is less than V_{in}, the output of the comparator, V_{comp}, is a low logic level. When V_{out} is greater than V_{in}, the output of the comparator is a high level.

The ADC performs a function opposite of the DAC; it converts an analog signal into digital form for processing by the microcomputer.

**Figure 4-18.
Analog-to-Digital
Converter**

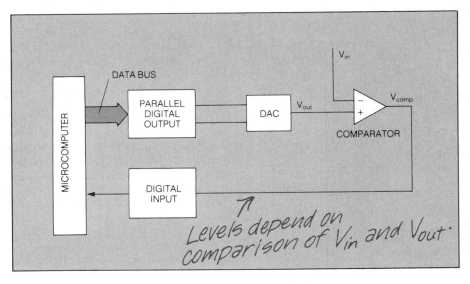

As soon as the binary number generated by the microcomputer causes V_{out} from the DAC to be greater than V_{in}, the comparator output goes high and stops the microcomputer from changing the binary number input further. The binary number is used by the microcomputer as the equivalent of the analog input voltage V_{in}. The microcomputer then resets and starts the binary number generation again to make another match to the V_{in} voltage. In this manner, a binary number, equivalent to a V_{in} analog voltage, is produced at a selected sampling rate. The output of the comparator is fed back to the microcomputer through a digital input.

Sampling

The accuracy of the DAC increases as the rate of sampling of the input signal increases.

The designer determines how quickly the microcomputer must change the DAC voltage to accurately follow the analog signal. *Figure 4-19* shows a sine wave analog signal and some digital approximations with various sampling rates. Notice that *Figure 4-19a* with 13 samples per sine wave cycle follows the sinusoid much closer than *Figure 4-19b* which only samples twice in a cycle. When the sampling rate is less than two as in *Figure 4-19c*, the staircase output doesn't follow well at all. This is because the computer didn't change the DAC input often enough to produce an output signal that closely approximates the desired signal.

Figure 4-19.
Analog Output Voltage
Versus Sampling Rate

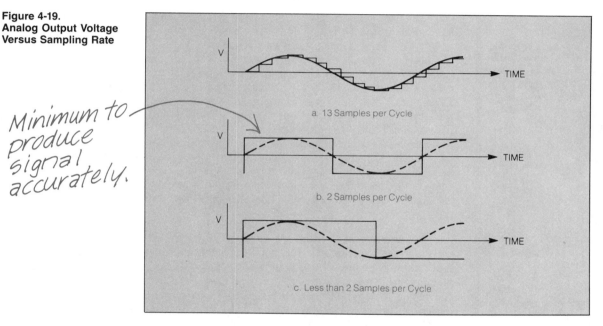

a. 13 Samples per Cycle

Minimum to produce signal accurately.

b. 2 Samples per Cycle

c. Less than 2 Samples per Cycle

The input sampling theorem states that an input signal must be sampled at least twice per cycle to be minimally accurate.

An engineer named H. Nyquist studied the sampling rate problem and determined that in order to reproduce a sinusoidal signal properly, the *signal must be sampled at least twice per cycle*. Of course, more samples per cycle is better, but two samples per cycle is the minimum that can be adequate. This result is known as the *Nyquist sampling theorem*. It can be applied to non-sinusoidal signals through the use of the frequency response methods of Chapter 2. In Chapter 2, it was shown that signals that were not sine waves could be generated by adding together a combination of different frequency sinusoids. To apply Nyquist's theorem to signals such as these, the designer selects the highest frequency sinusoid that he wants to reproduce and samples the signal at a rate which is at least twice the frequency of that sinusoid. This sampling frequency limit is called the Nyquist rate and is very important when designing digital control systems that uses analog signals.

Polling

Analog-to-digital converters are available that perform everything by themselves. The microcomputer simply tells them when to make a conversion and then waits until the conversion is done. ADCs require anywhere from a few millionths of a second to a second to complete the conversion.

Conversion time can be a serious limitation when slow converters are used. The microcomputer cannot afford to waste time waiting while the converter works. This is especially true when the microcomputer is used to control and monitor many systems at the same time. Instead of waiting for the ADC to finish, the computer could be off running another part of the program and come back only when the conversion is done. But how can the computer know when the conversion is finished?

Polling is used in some microcomputers to periodically check the ADC interface rather than waiting in an idle state for the ADC to do the conversion.

One way of doing this is for the microcomputer to periodically check the interface while it is running another part of the program. This method is called *polling*. A program subroutine is included in the main program that is called whenever an ADC interface has been called and is being used. This usually consists of a few lines of assembly language code which checks to see if the interface is done and collects the result when it is finished. When the polling subroutine determines that the ADC is finished, the main program continues without using the polling subroutine until the ADC interface is called again. The problem with such a scheme is that the polling routine may be called many times before the interface is finished. This can waste the computer's time and slow it down. Therefore, an evaluation must be made in certain systems to determine if polling is worthwhile.

Interrupts

Interrupts cause the CPU to jump to a specific location in the program. By signaling the microprocessor for service only when needed, interrupts are more efficient than polling.

An efficient alternative to polling uses control circuitry called an *interrupt*. An interrupt does just what the name implies; it tells the computer that something wants its attention. A slow analog-to-digital converter, for instance, could use an interrupt line to tell the processor when it is finished converting. When an interrupt occurs, the processor automatically jumps to a designated program location and executes the interrupt service subroutine. For the ADC, this would be a subroutine to read in the conversion result. When the interrupt subroutine is done, the computer returns to the place it left off in the program as if nothing had happened. (Recall the previous discussion on the jump-to-subroutine instruction.) Interrupts reduce the amount of time the computer spends dealing with the various peripheral devices.

Another important use for interrupts is in time keeping. Suppose we are controlling a system that requires things to be done at particular times. For instance, sampling an analog signal is a timed process. A special component called a timer could be used. A timer is a device which works like a digital watch. A square-wave clock signal is counted in counter registers like the one discussed in Chapter 3. The timer can be programmed to turn on the interrupt line when it reaches a certain count and then reset itself (start over). It may be inside the CPU itself or be contained in peripheral devices in the microcomputer system.

Such a technique could be used to output a new number to a digital-to-analog converter at regular intervals. The microcomputer simply programs the timer for the desired amount of time by presetting the counter to some starting value other than zero. Each time the timer counts out the programmed number of pulses, it interrupts the computer. The interrupt service subroutine then gets the new binary number which has been put into memory by the microcomputer and transfers this number to the DAC data latches at the input to the DAC.

Vectored Interrupts

All of the interrupt activity is completely invisible to the program that gets interrupted. In other words, the interrupted program doesn't know it was interrupted because its execution continues without program modification with minimum delay. Interrupts allow the computer to handle two or more things almost at once. In some systems, one interrupt line may be used by more than one device. For instance, two or more analog-to-digital converters may use the same interrupt line to indicate when either is ready. In this case, the computer doesn't know which device caused the interrupt. The computer could poll all the devices each time an interrupt occurs to see which one needs service, but as discussed, polling may waste time. A better way is to use vectored interrupts.

Vectored interrupts tell the CPU which specific device needs service, and also may implement a priority of service scheme. Vectored interrupts allow a microcomputer to handle a number of different tasks quickly.

A vector is something that points to another thing. It may be a specific memory location that contains the address of the first instruction of a subroutine to service an interrupt. It may be a register that contains the same type address. In this specific case, an interrupt vector is a register that peripherals use to tell the processor which device interrupted it. When a peripheral causes an interrupt, it writes a code into the interrupt vector register so the processor can tell which device interrupted it by reading the code. The decoder for an interrupt vector usually includes circuitry that allows each device to be assigned a different interrupt priority. If two devices interrupt at the same time, the processor will service the most important one first.

The vectored interrupt enables the microcomputer to efficiently handle the peripheral devices connected to it and service the interrupts rapidly. Interrupts allow the processor to respond to things happening in peripheral devices without having to constantly monitor the interfaces. They enable the microcomputer to handle many different tasks and to keep track of all of them. A microcomputer system designed to use interrupts is called a real time computing system because it rapidly responds to peripherals as soon as requests occur. Such real time systems are used in digital instrumentation and control systems.

MICROCOMPUTERS IN INSTRUMENTATION

Microcomputers can be used in instrumentation systems to improve their performance and efficiency. Chapter 9 will discuss this further. Recall that instrumentation systems are made of three basic functional blocks; the sensor, signal processing, and the display or actuator. A microcomputer can be used to perform the signal processing and display formatting required to improve the accuracy, reliability, and readability of the measurement system. Most sensors, as will be discussed in Chapter 5, convert the physical quantity they are measuring into an analog voltage. The microcomputer performs the signal processing and display formatting by first reading the analog sensor voltage with an analog-to-digital converter. Once the analog voltage has been sampled, the computer can then perform a wide variety of functions on it.

Sensor Linearization

Microcomputers can convert the non-linear output voltage of some sensors into a linear voltage representation. The sensor output voltage is used to look up the corresponding linear value stored in a table.

Sensor linearization is one important signal processing function which is commonly performed by a microcomputer. An example of a non-linear sensor which can be linearized by computer is the thermocouple temperature probe. The output voltage for a typical sensor is graphed as a function of temperature in *Figure 4-20*. Notice that the transfer characteristic for this device is not a straight line. If this voltage were sent directly to an analog meter, the scaling on the meter dial would have to be adjusted to correct for the nonlinearity. A microcomputer could compensate for the sensor's nonlinearity by reading the sensor's output voltage and looking up in a table stored in computer memory the temperature that voltage represents. The table would have two entries, one for the sensor's voltage and one for the sensor's temperature. The voltage the microcomputer would send to the analog meter (through a DAC) would be an adjusted linearized voltage corresponding to the table's temperature entry. The temperature numbers in the table would be chosen so that they cause the indicator to point to the correct temperature on the dial. The analog display device can be a standard display with one type of linear meter dial. This type of table look-up linearization can be done very quickly and easily with a microcomputer.

Display Linearization

Another feature of a microcomputer signal processor is that it helps in display linearization. In the temperature sensor example, the temperature values in the look-up table were chosen so that they caused the correct voltage to be sent to the display device assuming the display was linear. However, if the display device also happens to be nonlinear, this nonlinearity also can be compensated by adjusting the temperature values in the table. The microcomputer can perform sensor and display linearization at the same time by using one look-up table. If either the sensor or display is changed by design, the look-up table can be easily changed as necessary.

Figure 4-20.
Sensor Output Voltage

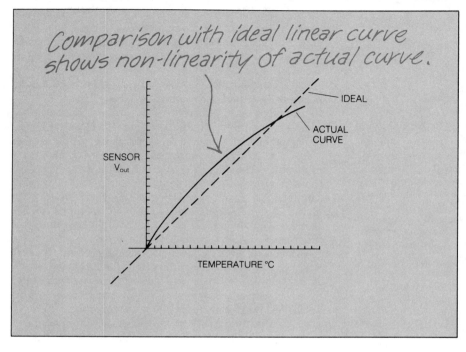

Digital Filters

Digital filters can be programmed on a microcomputer to reject specific types of signals while passing other signals.

A microcomputer signal processor can also be used to adjust the frequency response of sensors, actuators and display devices by performing signal filtering. Low pass filters pass low frequency signals, but reject high frequency signals. High pass filters do just the reverse; they pass high frequency signals and reject low frequency signals. Band pass filters pass midrange frequencies but reject both low and high frequencies. Analog filters use resistive, capacitive, and inductive components and sometimes analog amplifiers. Digital lowpass and and bandpass filters can be programmed on a microcomputer to perform basically the same function as their analog counterparts.

A digital lowpass filter could be used, for instance, to smooth the output of an automotive fuel level sensor. The fuel level sensor produces an electrical signal which is proportional to the height of the fuel in the center of the tank. The level at that point will change as fuel is consumed, but it also will change as the car slows, accelerates, turns corners, and hits bumps. The sensor's output voltage varies wildly because of fuel slosh even though the amount of fuel in the tank changes slowly. If that voltage is sent directly to the fuel gauge, the resulting variable indication will be hard to read.

The measurement can be made more readable and more meaningful by using a lowpass filter to smooth out the signal fluctuations to reduce the effects of sloshing. The lowpass filter can be implemented in a microcomputer by programming the computer to average the sensor signal over several seconds before sending it to the display. For instance, if the fuel level sensor signal is sampled once every second and we wish to average the signal over a period of 60 seconds, the computer saves only the latest 60 samples, averages them, and displays the average. When a new sample is taken, the oldest sample is discarded so only the 60 latest samples are kept. A new average can be computed and displayed each time a new sample is taken.

Digital filters, such as the averaging filter, are performed completely in software. They require no extra hardware once the signal has been read into the computer. This makes digital signal processing very attractive because the same computer can be used to process several different signals. Also, since digital filters require no extra hardware, the filters can be made much more complex with no increase in cost. In addition, the characteristics of the digital filter can be changed by changing the software. Changing the characteristics of an analog filter usually requires changing the hardware components. Another feature of digital filters (true for digital signal processing in general) is that they don't change with age or temperature as analog filters tend to. (The software is the same at 10°C as it is at 80°C, but the characteristics of analog hardware usually change with temperature.)

However, there are limitations also. The frequency range of digital filters is determined by the speed of the processor. The microcomputer must be able to sample each signal at or above the rate required by the Nyquist sampling theorem. It must also be fast enough to perform all of the signal averaging and linearization for each signal before the next sample is taken. This is an important limitation and the system designer must be certain that the computer is not overloaded by trying to make it do too many things too quickly.

Cost Versus Performance

Another important aspect of microcomputer signal processing is the trade off between system cost and performance. Microcomputer systems can cost several hundred dollars depending on their complexity. All of the instrumentation functions that have been discussed so far can be implemented using analog circuitry, and the cost of the microcomputer system could buy a lot of analog circuitry. Even though digital signal processing is generally more precise and versatile, these features may not be necessary in a particular application. Even though they come somewhat automatically with a microcomputer, a bargain is not a bargain if the microcomputer is not needed in the first place. As we will see, the justification for using a microcomputer in

Digital filters are performed completely through the use of software, thus their characteristics can be easily changed. Because digital filters require no extra hardware, they are low in cost.

automotive instrumentation lies in its ability to control overall functions of performance and emissions with the accuracy required, with reliability, and with cost effectiveness. Cost effectiveness, because, since the system is required for the control accuracy, many more functions are performed with the same computer hardware. Also, instrumentation changes can easily be made by changing the computer's software rather than modifying circuit boards or other hardware.

MICROCOMPUTERS IN CONTROL SYSTEMS

Microcomputers are able to input and output both digital and converted analog signals. With the proper software, they are capable of making decisions about those signals and can react to them quickly and precisely. These features make microcomputers ideal for controlling other digital or analog systems. Let's look at several of these general systems.

Closed Loop Control System

With proper software, a microcomputer can replace the error amplifier and control logic used in the closed loop control system.

Recall the basic closed loop control system block diagram of Chapter 2. The error amplifier compares the command input with the plant output and sends the error signal to the control logic. The control logic uses the error signal to generate a plant control signal which causes the plant to react with a new output so that the error signal will be reduced toward zero. The control logic is designed so that the plant's output follows or tracks the command input.

A microcomputer can replace the error amplifier and the control logic. The computer can compare command input and plant output and perform the computation required to generate a control signal.

Limit Cycle Controller

The limit cycle controller discussed in Chapter 2 can easily be implemented with a microcomputer. Recall that the limit cycle controller controls the plant output so that it falls somewhere between an upper and lower limit, preferably so that its average value is equal to the command input. The controller must read in the command input and the plant output and decide what control signal to send to the plant based on those signals alone.

Using a microcomputer, the upper and lower limit can be determined from the command input by using a look-up table similar to the technique used for linearizing the temperature sensor in an earlier discussion. The plant output is compared against these two limits. If the plant output is above the higher limit or below the lower limit, the microcomputer outputs the appropriate on/off signal to the plant to bring the output back between the two limits.

The microcomputer can also be programmed to monitor itself and to keep track of how well it is controlling the plant output. The microcomputer can keep a running average of the command and output signals. By comparing these averages, it can tell if it is consistently controlling above or below the average command input. This information can be used by the microcomputer to adjust the upper and lower limits in the look-up table to bring the averages closer. Many such schemes can be easily implemented with a computer, but would be difficult with analog circuitry.

Other types of control logic can also be designed using microcomputers. The complexity of the control scheme can be greatly increased with digital logic because it only costs programming time, which is a one-time cost. No extra hardware (which would be an extra cost on each unit produced) is required to alter or improve the control scheme in systems using microcomputers.

Multivariable Systems

With the appropriate control scheme program, microcomputers have the ability to sample and control multiple inputs and outputs independently. This type of control is much more difficult to design when using analog circuitry.

A very important feature of microcomputer control logic is the ability to control multiple systems independently or systems with multiple inputs and outputs. The automotive applications for microcomputer control involves both of these types of so called "multivariable" systems. For instance, the automobile engine has several inputs (such as air/fuel ratio, throttle angle, spark timing, etc.) and several outputs (torque, speed, exhaust gases, etc.). All of the outputs must be controlled simultaneously because some inputs affect more than one output. These types of controllers can be very complicated and are difficult to implement in analog fashion. The increased complexity of these systems doesn't affect the complexity or cost of the microcomputer system once the microcomputer is chosen of a size to do the task. It only affects the task of programming the appropriate control scheme into the microcomputer.

SUMMARY

In this chapter, the discussion centered on what microcomputers are capable of doing, how they do it, and what they can do in instrumentation and control systems. General microcomputer operations and some of the peripheral interfaces which enable them to interact with other digital and analog systems were discussed, along with interrupts which allow the computer to handle these peripherals and their interfaces at high speed without wasting time. System applications which demonstrated some basic ideas and techniques of instrumentation and control were examined.

After a chapter on sensors and actuators, we will deal more specifically with particular microcomputer automotive instrumentation and control systems to show how these systems are used in the automobile to control the engine and drive train, and many auxiliary functions besides.

Quiz for Chapter 4

1. What does a microcomputer do best?
   a. add, subtract, multiply and divide large numbers
   b. overcome temperature drift
   c. eliminate noise problems
   d. none of the above
   e. all of the above

2. What does a microcomputer use to interface with other systems?
   a. parallel interface
   b. analog to digital converter
   c. digital to analog converter
   d. all of the above

3. Which control line do peripherals use to get the computer's attention?
   a. power line
   b. read/write line
   c. interrupt line
   d. clock line

4. What is a data bus?
   a. a set of wires which carry bits to or from the processor and memory or peripherals
   b. a large yellow vehicle for carrying datas
   c. a bus carrying addresses
   d. a set of wires for control signals

5. What are computers used for in instrumentation systems?
   a. signal processing
   b. sensor, actuator and display linearization
   c. display formatting
   d. filtering
   e. all of the above

6. Which type of digital filter was discussed by example in this chapter?
   a. analog
   b. highpass
   c. lowpass
   d. bandpass

7. What advantages does digital signal processing have over analog signal processing?
   a. digital is more precise
   b. digital doesn't drift with time and temperature
   c. the same digital hardware can be used in many filters
   d. all of the above

8. What advantages does analog signal processing have over digital signal processing
   a. analog is always less expensive
   b. the same analog hardware can be used for many filters
   c. analog is sometimes less expensive
   d. high frequency signals can only be filtered with analog filters

9. What type of memory is used to permanently store programs?
   a. RAM
   b. ROM
   c. MAP
   d. RPM

10. What type of memory is used to temporarily store data and variables?
    a. RAM
    b. ROM
    c. MAP
    d. RPM

11. What distinguishes a computer from a fancy calculator?
    a. add, subtract, multiply and divide
    b. stored program
    c. the calculators can read paper tape
    d. digital circuits

12. What part of the computer does the arithmetic and logic functions?
    a. peripherals
    b. memory
    c. CPU
    d. address bus

13. Which computer register is the main
work register?
  a. program counter
  b. stack pointer
  c. condition code register
  d. accumulator

14. A short initialization program is
called what kind of program?
  a. subroutine
  b. boot program
  c. main program
  d. branch

15. Which register keeps track of
program steps?
  a. program counter
  b. stack pointer
  c. condition code register
  d. accumulator

16. A programmer uses what type of
statements in an assembly language
program?
  a. op codes
  b. mnemonics
  c. machine code

17. Most microcomputers use how many
bits to address memory?
  a. 1
  b. 16
  c. 4
  d. 6

18. Most automotive microcomputers
use how many bits in arithmetic?
  a. 1
  b. 6
  c. 4
  d. 8

19. Which of the following is a short
program which ends with an RTS
instruction?
  a. main program
  b. interrupt
  c. boot
  d. subroutine

20. Why are interrupts used?
  a. eliminate polling
  b. avoid wasting time
  c. smooth peripheral controls
  d. all of the above

Sensors and Actuators

ABOUT THIS CHAPTER

In this chapter, the theory and operation of two vital components of automotive control and instrumentation systems will be explained; the sensor and the actuator. The sensor is an input device that gives the system information to be used to determine an action. The actuator is an output device in an electronic control system that performs some action in response to an electrical input. The output device for instrumentation systems is called a display. The emphasis in this chapter will be on actuators that perform an action. Displays will be discussed in Chapter 9. In this examination, some of the shortcomings of existing sensor performance will be discovered and the needs for future development or invention will be apparent. Hopefully, understanding the basic concepts and the needs will promote the thinking necessary to significantly improve the performance of an existing device or even invent a new automotive sensor. The future is wide open for such invention!

AUTOMOTIVE CONTROL SYSTEM APPLICATIONS

Sensors and actuators play a critical role in determining automotive control system performance.

In control system applications, sensors and actuators, in many cases, are the critical components for determining system performance. This is especially true for automotive control system applications. The availability of appropriate sensors and actuators dictates the design of the control system and the type of function it can perform.

The sensors and actuators which are available to a control system designer are not always what the designer wants because the required device is not manufactured. For this reason, special signal processors or interface circuits many times must be designed to adapt to an available sensor or actuator, or the control system controller is designed in a specific way to fit available sensors or actuators. However, for automotive control systems, because of their large potential quantity, it has been worthwhile to develop a sensor for a particular application, even though it often has taken a long and expensive research project to do so.

OVERVIEW

Signal flow in a monitoring system is from a sensor to a signal processor, and finally to a display or actuator.

To understand how sensors and actuators are used in a system, refer to *Figure 5-1*. As pointed out in Chapter 2, q_0 is a physical quantity. The sensor converts the physical quantity, q_0, into an electrical signal, q_1, that represents the value of the physical quantity being measured. The electronic signal processor operates on q_1, the electrical signal from the sensor, and generates another electrical signal, q_2, in a form suitable to operate an actuator, if the system under consideration is a control system, or to operate a display device, if the system is an instrumentation application. q_3 is the output action of the actuator or the readable quantity of the display. In *Figure 5-1*, we aren't concerned with whether the control system is operating in an open or closed loop. Rather, we wish to emphasize the signal flow from the sensor through some kind of processor to the display or actuator.

**Figure 5-1.
Instrumentation Block
Diagram**

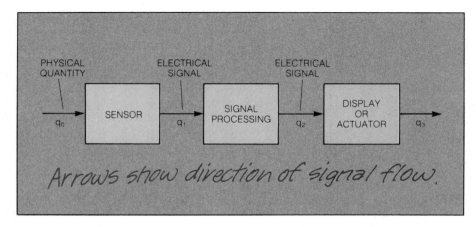

In a control application, it is necessary to measure one or more physical quantities (to be more general we'll call them variables). The control system uses the value of these variables to determine the status of the system being controlled. The electronic control system can operate only with signals in electrical form; therefore, a sensor must be used to convert the physical variable being measured into a suitable electrical signal. Actuators usually require some form of mechanical action; therefore, the output of the signal processor is a reconversion of the electrical signal into suitable mechanical action.

VARIABLES TO BE MEASURED

Figure 5-2 shows a simplified electronic control system for an internal combustion automotive engine. The basic inputs to the engine are air and fuel and the basic outputs are the mechanical drive power and the exhaust emissions. Maximum efficiency of the conversion of fuel to drive power is desired while the exhaust emissions of burned by-products are maintained within desirable limits. The air/fuel ratio is a key control parameter. Sensors measure the physical variables and feed electrical signals through signal processors to the controller. The controller generates the electrical outputs which operate the actuators to control the engine performance.

Many variables must be measured. These include:

1) coolant temperature
2) inlet air temperature
3) manifold absolute pressure
4) atmospheric absolute pressure
5) crankshaft angular position
6) engine angular speed (RPM)
7) exhaust gas oxygen concentration

At least one sensor that has acceptable performance for measuring each of these variables is presently available.

An electronic control system for an automotive engine must deal with variables such as engine speed, various temperatures and pressures, crankshaft position, and exhaust gas oxygen concentration.

**Figure 5-2.
Simplified Electronic
Engine Control System**

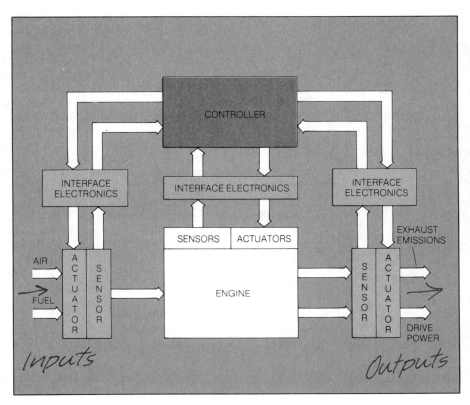

There are other variables which the control system designer would like to measure directly, but a cost-effective sensor does not presently exist. These variables include:

1) engine output (brake torque)
2) in-cylinder pressure
3) mass flow rate of air into each cylinder

Often the variable which the control system designer wants to measure can be obtained only indirectly by sensing a closely related variable. This will be illustrated with several automotive examples. It will be shown that it is sometimes necessary to measure several variables and perform complex mathematical operations just to obtain a measurement of a desired variable.

ANALYSIS OF INTAKE MANIFOLD PRESSURE

The MAP sensor output voltage is proportional to the average pressure within the intake manifold.

The air and fuel mixture enters the engine in a number of ways but usually always through the intake manifold. The intake manifold is a series of channels and passages that direct the air and fuel mixture to the cylinders. One very important engine variable associated with the intake manifold is the manifold absolute pressure (MAP). The sensor that measures this pressure is the manifold absolute pressure sensor—the MAP sensor. This sensor develops a voltage which is approximately proportional to the average value of intake manifold pressure. In Chapter 6, we'll see that the MAP sensor voltage has many applications in an engine electronic control system.

Figure 5-3 is a very simplified sketch of an intake manifold. In this simplified sketch, the engine is viewed as an air pump drawing air into the intake manifold. Whenever the engine is not running, no air is being pumped and the intake manifold absolute pressure is at atmospheric pressure. This is the highest intake manifold absolute pressure for an unsupercharged engine (a supercharged engine has an external air pump called a supercharger). The throttle plate shown is normally in the carburetor. When the engine is running, the air flow is impeded by the partially closed throttle plate in the carburetor. This reduces the pressure in the intake manifold so it is lower than atmospheric pressure; therefore, a partial vacuum exists in the intake manifold.

The manifold absolute pressure varies from near atmospheric pressure when the throttle plate is fully opened to near zero pressure when the throttle plate is closed.

If the engine were a perfect air pump and if the throttle plate were tightly closed, a perfect vacuum could be created in the intake manifold. A perfect vacuum corresponds to zero absolute pressure. However, the engine is not a perfect pump and some air always leaks past the throttle plate. (In fact, some air must get past the throttle plate or the engine can't run.) Therefore, the intake manifold absolute pressure is slightly above absolute zero when the throttle plate is closed and the engine is running. At the other extreme, when the engine is running and the throttle plate is wide open, the manifold pressure is nearly equal to the atmospheric pressure. Thus, when the engine is running, the intake manifold pressure varies from nearly zero with the throttle plate closed to nearly atmospheric with the throttle plate wide open.

Figure 5-3.
Simplified Intake System

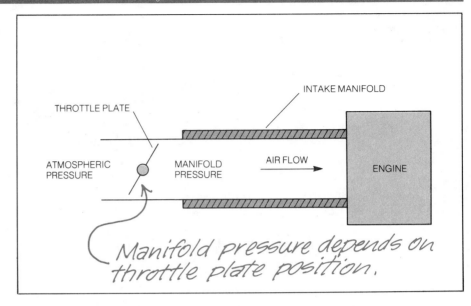

Manifold pressure depends on throttle plate position.

Manifold pressure fluctuates because of the pumping action of the individual cylinders. The MAP sensor filters these fluctuations so its output represents the average pressure.

The variations in intake manifold pressure as the throttle plate position is held constant also must be considered. The manifold pressure fluctuates rapidly because of the individual pumping action of the several cylinders. Each cylinder begins drawing in air when its intake valve opens and its piston begins the downward motion after top dead center (TDC). Manifold pressure decreases during this time. The drawing in of air for this cylinder ends when bottom dead center (BDC) is reached and the intake valve closes. The manifold pressure begins to increase until another cylinder begins to draw in air, then the pressure decreases again. Therefore, the manifold pressure fluctuates during the stroke of each cylinder and as pumping is switched from one cylinder to the next.

Each cylinder contributes to the pumping action every second crankshaft revolution. For an N cylinder engine, the frequency, f_p, in cycles per second, of the manifold pressure fluctuation for an engine running at a certain RPM is given by:

$$f_p = \frac{N \times RPM}{120}$$

The MAP fluctuations for 1200 and 2400 RPM are shown in *Figure 5-4*.

For a control system application, only the average manifold pressure is required. Therefore, the manifold pressure measurement method should filter out the pressure fluctuations at frequency fp and measure only the average pressure. One way to achieve this filtering is to connect the MAP sensor to the intake manifold through a very small diameter tube. The rapid fluctuations in pressure don't pass through this tube, but the average pressure does. The MAP sensor output voltage then corresponds only to the average manifold pressure.

Figure 5-4.
Intake Manifold Pressure
Fluctuations

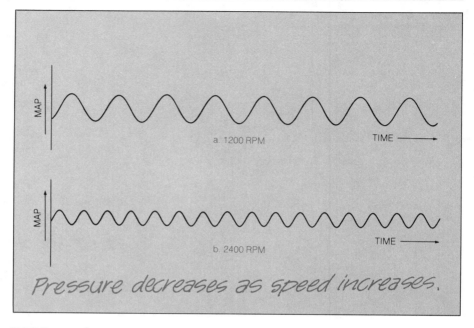

Pressure decreases as speed increases.

MAP Sensor Concepts

Several MAP sensor configurations have been used in automotive applications. The earliest sensors were derived from aerospace instrumentation concepts, but these proved more expensive than desirable for automotive applications and have been replaced with more cost effective designs. Actually, research is continuing (and will likely continue) for lower cost and better performing MAP sensors. Several of the concepts used for sensors of manifold absolute pressure will be discussed.

It is interesting to note that none of the MAP sensors in use measure manifold pressure directly, but instead measure the displacement of a diaphragm which is deflected by manifold pressure. The details of the diaphragm displacement and the measurement of this displacement vary from one configuration to another.

Aneroid MAP Sensor

In an aneroid MAP sensor, a pair of diaphrams that respond to pressure variations are connected to the moving core in a special type of transformer. As the core moves, the output voltage of the transformer varies.

One of the earliest MAP sensor configurations used in automotive applications is illustrated in *Figure 5-5*. This sensor actually uses a pair of diaphragms which are welded together under vacuum to form an aneroid chamber. The aneroid chamber is placed in a sealed housing which is connected by a small diameter tube to the intake manifold so that the pressure inside the sealed housing is the average manifold absolute pressure. The manifold pressure deforms the aneroid such that increasing manifold pressure compresses the aneroid. The aneroid is designed such that the displacement is almost perfectly linear with manifold pressure.

**Figure 5-5.
Typical Aneroid MAP
Sensor**

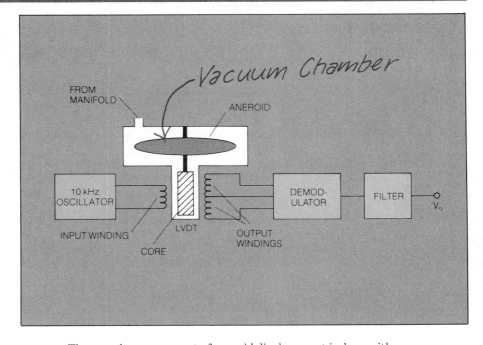

The actual measurement of aneroid displacement is done with a sensor called a linear variable differential transformer (LVDT). The LVDT has a movable core and the coupling between the input and output transformer windings varies with core position. A 10 kHz signal is applied to the input winding. The output windings of the LVDT are balanced so that the output voltage of one winding is equal to the other with the core at its center position. However, they are connected so they cancel each other and the net output voltage is zero. As the core is displaced from this mid-position by manifold pressure, the output voltage of one winding is greater than the other, so the net output voltage varies in proportion to the amount of movement of the core.

Use of the aneroid MAP sensor is being discontinued in favor of less expensive sensor designs.

The electronic signal processing (demodulator and filter) produces a dc voltage which is proportional to the manifold absolute pressure, as shown in *Figure 5-6*. Unfortunately, this sensor is relatively expensive for large-scale automotive application and its use is being discontinued. It is being replaced with more cost-effective designs.

Figure 5-6.
V_o Versus MAP for MAP Sensor

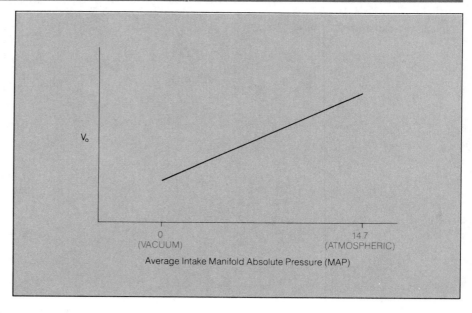

V_o

0	14.7
(VACUUM)	(ATMOSPHERIC)

Average Intake Manifold Absolute Pressure (MAP)

Strain Gauge MAP Sensor

In the strain gauge MAP sensor, manifold pressure applied to the diaphragm causes a resistance change within the semiconductor material that corresponds to the manifold pressure.

One relatively inexpensive MAP sensor configuration is the silicon diaphragm diffused strain gauge sensor (SCSG) shown in *Figure 5-7*. This sensor uses a silicon chip which is approximately 3 millimeters square. Along the outer edges, the chip is approximately 250 micrometers (1 micrometer = 1 millionth of a meter) thick but the center area is only 25 micrometers thick to form a diaphragm. The edge of the chip is sealed to a pyrex plate under vacuum thereby forming a vacuum chamber between the plate and the center area of the silicon chip.

A set of sensing resistors are formed around the edge of this chamber as indicated in *Figure 5-7*. The resistors are formed by diffusing a 'doping impurity' into the silicon[1]. External connections to these resistors are made through wires connected to the metal bonding pads.

This entire assembly is placed in a sealed housing which is connected to the intake manifold by a small diameter tube. Manifold pressure applied to the diaphragm causes it to deflect. The resistance of the sensing resistors changes in proportion to the applied manifold pressure by a phenomenon which is known as piezo-resistivity. Piezo-resistivity occurs in certain semiconductors so that the actual resistivity (a property of the material) changes in proportion to the strain (fractional change in length).

[1] *Understanding Solid-State Electronics*, Engineering Staff of Texas Instruments, 1972, Chapter 7, Texas Instruments Incorporated.

**Figure 5-7.
Typical Silicon-
Diaphragm Strain Gauge
MAP Sensor**

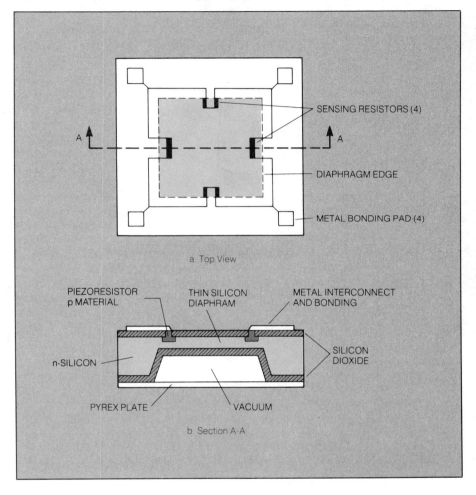

SENSING RESISTORS (4)

DIAPHRAGM EDGE

METAL BONDING PAD (4)

a. Top View

PIEZORESISTOR
p MATERIAL

THIN SILICON
DIAPHRAM

METAL INTERCONNECT
AND BONDING

SILICON
DIOXIDE

n-SILICON

PYREX PLATE   VACUUM

b. Section A-A

The resistors in the strain guage MAP sensor are connected in a Wheatstone bridge circuit. Output voltage of the circuit varies as the resistance varies in response to manifold pressure variations.

An electrical signal which is proportional to the manifold pressure is obtained by connecting the resistors in a circuit called a "Wheatstone bridge" as shown in the schematic diagram of *Figure 5-8a*. The voltage regulator holds a constant dc voltage across the bridge. The resistors diffused into the diaphragm are denoted R_1, R_2, R_3 and R_4 in *Figure 5-8a*. When there is no strain on the diaphragm, all four resistances are equal, the bridge is balanced, and the voltage between points A and B is zero. When manifold pressure changes, it causes these resistances to change in such a way that R_1 and R_3 increase by an amount which is proportional to pressure and, at the same time, R_2 and R_4 decrease by an identical amount. This unbalances the bridge and a net difference voltage is present between points A and B. The differential amplifier generates an output voltage proportional to the difference between the two input voltages, as shown in *Figure 5-8b*.

**Figure 5-8.
Circuit Using a Strain
Gauge MAP Sensor**

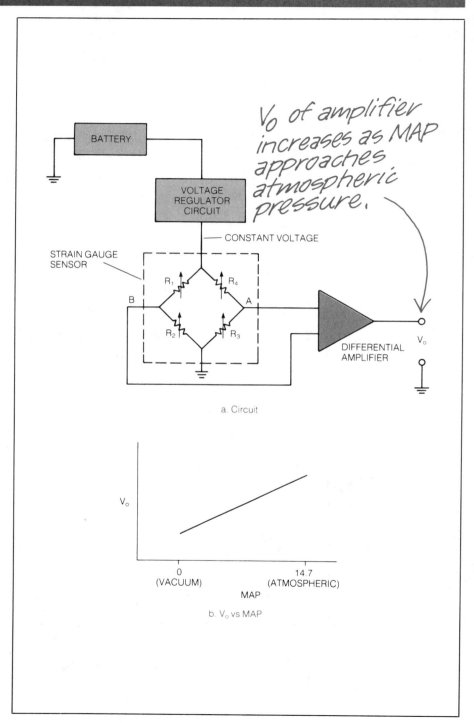

V_o of amplifier increases as MAP approaches atmospheric pressure.

a. Circuit

b. V_o vs MAP

Capacitor-Capsule MAP Sensor

In the capacitor capsule MAP sensor, two flexible metal plates separated by an insulating spacer and air form a capacitor. Variations in manifold pressure vary the distance between plates. The resultant change in capacitance indicates absolute manifold pressure.

Another interesting MAP sensor configuration is the capacitor-capsule MAP sensor shown in *Figure 5-9.* A film electrode is deposited on the inside face of each of the two alumina plates and a connecting lead is extended for external connections. The plates are sealed together with an insulating hollow cylindrical spacer (i.e. washer shaped) between to form an aneroid chamber. The capacitor capsule is placed inside a · sealed housing which is connected to manifold pressure by a small diameter tube. The film electrodes face one another on the inside of this aneroid and form a parallel plate capacitor. The alumina plates are flexible so that they deflect inward under the influence of manifold pressure. The deflection of these plates causes the distance between the electrodes to change in response to manifold pressure.

**Figure 5-9.
Capacitor-Capsule MAP
Sensor**

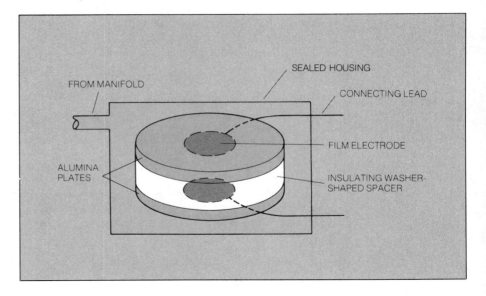

Recall that the capacitance, C, of such a capacitor is given approximately by:

$$C = \frac{\epsilon_0 A}{d}$$

where

ϵ_0 = dielectric constant for air,
A = area of film electrodes, and
d = distance between electrodes.

The manifold pressure causes the distance, d, to decrease as pressure increases; therefore, the capacitance increases as pressure increases as indicated in *Figure 5-10*.

**Figure 5-10.
Variation in MAP Sensor
Capacitance with
Manifold Pressure**

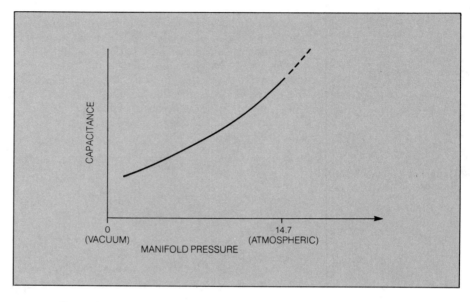

The change in capacitance can be used to generate a voltage by placing the sensor in a resonant circuit. Changes in the sensor's capacitance causes corresponding changes in the output voltage.

Several methods can be used to generate a voltage in proportion to the change in capacitance so that manifold pressure can be measured. One relatively simple and inexpensive scheme involves connecting the sensor in a series resonant circuit as shown in *Figure 5-11*. In this circuit, the inductor and capacitive sensor form a series resonant circuit whose resonant frequency, f_r, is given by:

$$f_r = \frac{1}{2\pi \sqrt{LC}}$$

where

L is inductance in henrys
C is capacitance in farads
$\pi = 3.1416$

The oscillator frequency is tuned to the circuit's resonant frequency for atmospheric absolute manifold pressure. At resonance, the voltage across the inductor and the voltage across the capacitor are equal, but of opposite phase, and cancel each other so that the voltage across the resistor is the applied voltage. Therefore, the output, V_o, of the phase detector is zero volts because the voltage across the resistance, V_R, is in phase with the reference phase. As manifold pressure varies the capacitance in the circuit, the resonant frequency is changed. Since the oscillator frequency stays the same, the phase of the voltage across R relative to the reference phase varies sharply. The phase detector detects the change in phase and produces a voltage, V_o, which is

Figure 5-11.
Signal Processing for
Capacitor-Capsule MAP
Sensor

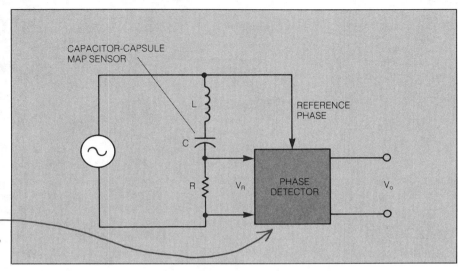

Phase detector compares phase of oscillator to reference phase.

Crankshaft angular position is an important variable in automotive control systems, particularly for controling ignition timing and fuel injection timing.

proportional to the change in phase. Therefore, the output voltage of the phase detector is proportional to the manifold pressure.

ENGINE CRANKSHAFT ANGULAR POSITION SENSOR

Besides pressure, the position of shafts, valves, levers, etc. must be sensed for automotive control systems. Let's look at the concepts used by several sensors for important position variables. A significant variable which must be sensed in an automotive control system is crankshaft angular position. Imagine the engine as viewed from the rear as shown in *Figure 5-12*. On the rear of the crankshaft is a large, heavy, circular steel disk called the flywheel which is connected to and rotates with the crankshaft. Let's mark a point on the flywheel as shown in *Figure 5-12* and draw a line through this point and the axis of rotation. Let's draw another line through the axis of rotation parallel to the horizontal center line of the engine. This line is simply for a reference line. The crankshaft angular position is the angle between the reference line and the mark on the flywheel.

Imagine that the flywheel is rotated so that the mark is directly on the reference line. This is an angular position of zero degrees and, for our purposes, assume that this angular position corresponds to the No. 1 cylinder at TDC (top dead center). As the crankshaft rotates, this angle increases from zero to 360° in one revolution. However, one full engine cycle from intake through exhaust requires two complete revolutions of the crankshaft. That is, one complete engine cycle corresponds to the crankshaft angular position going from zero to 720°. During each cycle, it is important to measure the crankshaft position with reference to TDC for each cylinder. This information is used by the electronic engine controller to set ignition timing and, in some cases, to adjust the fuel control system parameters.

Figure 5-12.
Engine Crankshaft
Angular Position
Measurement

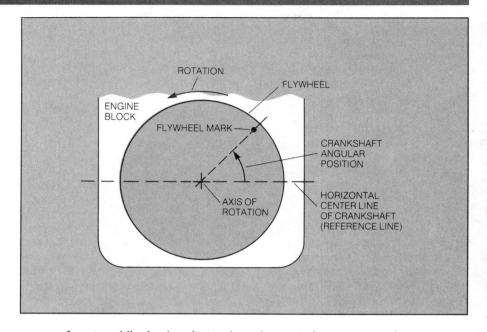

Crankshaft angular position can be sensed directly at the crankshaft, or at the camshaft, since the camshaft rotates at exactly one-half the speed of the crankshaft.

In automobiles having electronic engine control systems, angular position can be sensed on the crankshaft directly or on the camshaft. Recall that the piston drives the crankshaft directly, while the valves and the distributor for the spark ignition are driven from the camshaft. The camshaft is driven from the crankshaft through a 1:2 reduction drive train, which can be gears, belt or chain. Therefore, the camshaft rotational speed is one-half that of the crankshaft so the camshaft angular position goes from zero to 360° for one complete engine cycle. Either of these sensing locations have been used in one electronic control system or another. Although the crankshaft location is superior for accuracy because of torsional and gear backlash errors in the camshaft drive train, many production systems locate this sensor in the distributor where it measures camshaft position. At the present time, there appears to be a trend toward measuring crankshaft position directly. Examples of both sensor configurations for engine angular position measurement will be presented.

It is desirable to measure engine angular position with a non-contacting sensor to avoid mechanical wear and corresponding changes in accuracy of the measurement. The two most common methods for non-contact coupling to a rotating shaft employ magnetic fields or optics. Let's consider the concepts used for magnetically coupled sensors.

Magnetic Reluctance Position Sensor

In the magnetic reluctance position sensor, a coil wrapped around the magnet senses the changing intensity of the magnetic field as the tabs of a ferrous disk pass between the poles of the magnet.

One engine sensor configuration which measures crankshaft position directly is illustrated in *Figure 5-13*. This sensor consists of a permanent magnet with a coil of wire wound around it. A steel disk which is mounted on the crankshaft (usually in front of the engine) has tabs that pass between the pole pieces of this magnet. In *Figure 5-13*, the steel disk has four protruding tabs which are appropriate for an 8-cylinder engine. The passage of each tab corresponds to the TDC position of a cylinder on its power stroke.

**Figure 5-13.
Magnetic Reluctance
Crankshaft Position
Sensor**

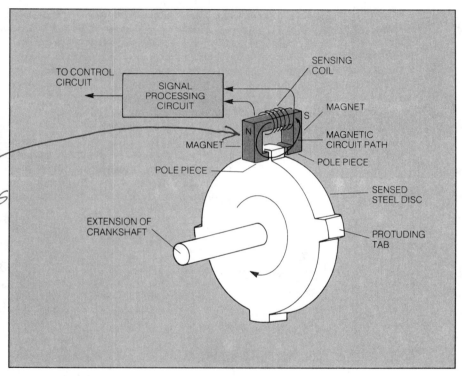

Change in magnetic flux produces changes in sense coil output voltage.

This sensor is of the magnetic-reluctance type and is based upon the concept of a "magnetic circuit". A magnetic circuit is a closed path through a magnetic material (i.e. iron, cobalt, nickel or man-made magnetic material called ferrite). In the case of the sensor in *Figure 5-13*, the magnetic circuit is the closed path through the magnet material and across the gap between the pole pieces.

The magnetic field in a magnetic circuit is described by a pair of field quantities which can be compared to the voltage and current of an ordinary electric circuit. One of these quantities is called the magnetic field intensity. It exerts a force similar to the voltage of a battery.

The response of the magnetic circuit to the magnetic field intensity is described by the second quantity which is called magnetic flux. A line of constant magnetic flux is a closed path through the magnetic material. The magnetic flux is similar to the current which flows when a resistor is connected across a battery forming a closed electric circuit.

As we shall see, the voltage generated by the reluctance sensor is determined by the strength of this magnetic flux. The strength of the magnetic flux is, in turn, determined by the reluctance of the magnetic circuit. Reluctance is to a magnetic circuit what resistance is to an electric circuit.

The path for the magnetic flux of the reluctance sensor is illustrated in *Figure 5-14*. The reluctance of a magnetic circuit is inversely proportional to the magnetic permeability of the material along the path. The magnetic permeability of steel is roughly a few thousand times larger than air; therefore, the reluctance of steel is much lower than air. Note that when one of the tabs of the steel disk is located between the pole pieces of the magnet, a large part of the gap between the pole pieces is filled by the steel. Since the steel has a lower reluctance than air, the "flow" of magnetic flux increases to a relatively large value.

On the other hand, when a tab is not between the magnet pole pieces, the gap is filled by air only. This creates a high reluctance circuit for which the magnetic flux is relatively small. Thus, the magnitude of the magnetic flux which "flows" through the "magnetic circuit" depends upon the position of the tab which, in turn, depends on the crankshaft angular position.

> The voltage generated by the magnetic reluctance position sensor is determined by the strength of the magnetic flux. When a tab on the steel disk passes through the gap, the flow of the magnetic flux changes significantly.

**Figure 5-14.
Magnetic Circuit of the
Reluctance Sensor**

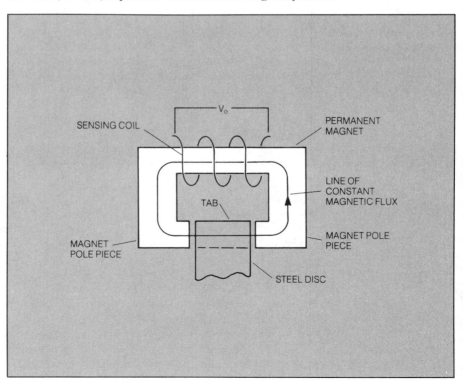

The magnetic flux is least when none of the tabs is near the magnet pole pieces. Then, as a tab begins to pass through the gap, the magnetic flux increases, reaches a maximum when the tab is exactly between the pole pieces, then decreases as the tab passes out of the pole piece region. In most control systems, the position of maximum magnetic flux corresponds to TDC for one of the cylinders.

A voltage, V_o, *is induced in the sensing coil by the change in magnetic flux which is proportional to the rate of change of the magnetic flux.* Since the magnetic flux must be changing to induce a voltage in the sensing coil, its output voltage is zero whenever the engine is not running regardless of the position of the crankshaft. *This is a serious disadvantage for this type sensor because the engine timing cannot be set statically.*

As shown in *Figure 5-15*, the coil voltage, V_o, begins to increase from zero as a tab begins to pass between the pole pieces, reaches a maximum, then falls to zero when the tab is exactly between the pole pieces. (Note that although the value of magnetic flux is maximum at this point, *the rate of change of magnetic flux is zero*; therefore, the induced voltage in the sensing coil is zero.) Then it increases with the opposite polarity, reaches a maximum, and falls to zero as the tab passes out of the gap between the pole pieces. The coil voltage waveform shown in *Figure 5-15b* occurs each time a tab passes between the pole pieces. Thus, a voltage pulse having the waveform of *Figure 5-15b* occurs each time one of the cylinders reaches TDC on its power stroke.

The voltage induced in the sensing coil varies with the rate of change of the magnetic flux. When the tab is centered between the poles of the magnet, the voltage is zero because the flux is not changing.

**Figure 5-15.
Output Voltage
Waveform from the
Magnetic Reluctance
Crankshaft Position
Sensor Coil**

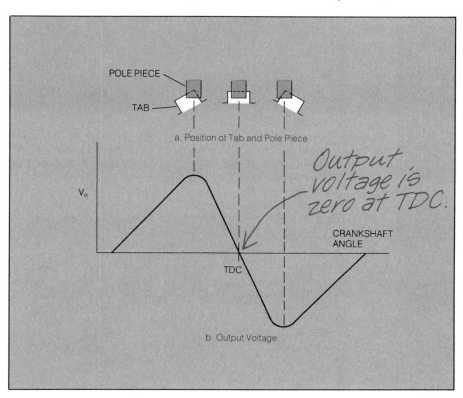

It should be noted that the number of tabs always will be half the number of cylinders for this crankshaft position sensor because it takes two crankshaft rotations for a complete engine cycle.

Engine Speed Sensor

Engine speed can be calculated in a number of ways. Digital circuits use counters and crankshaft sensors to calculate actual engine speed.

An engine speed sensor is needed to provide an input for the electronic controller for several functions. The position sensor discussed above can be used to measure engine speed. The reluctance sensor is used in this case as an example; however, any of the other position sensor techniques could be used as well. Refer to *Figure 5-13* and notice that the four tabs will pass through the sensing coil once for each crankshaft revolution. Therefore, if we count the pulses of voltage from the sensing coil in one minute and divide by four, we will know the engine speed in revolutions per minute (RPM).

This is easy to do with digital circuits. Precise timing circuits such as those used in digital watches can start a counter circuit which will count pulses until the timing circuit stops it. The counter can have the divide-by-4 function included in it or a separate divider circuit may be used. In many cases, the actual RPM sensor disk is mounted near the flywheel and has many more than four tabs and the counter does not actually count for a full minute before the speed is calculated, but the results are the same.

Ignition Timing Sensor

The notched position sensor uses an effect opposite that of the tab position sensor. As a notch in a rotating steel disk passes by a variable reluctance sensor, the decrease in magnetic flux generates a voltage pulse in the sensor coil.

For some types of electronic engine control, it is desirable to have a sensor for detecting a single reference point during one revolution of the crankshaft. One scheme for detecting this reference point uses the harmonic damper, a steel, disk-shaped device which is connected to the crankshaft at the end opposite the flywheel. The harmonic damper has a notch cut in its outer surface as shown in *Figure 5-16*.

A variable reluctance sensor is mounted on the engine block near the harmonic damper. The harmonic damper provides a relatively low reluctance path for the magnetic flux of the sensor because it is made of steel.

Whenever the notch aligns with the sensor axis, the reluctance of this magnetic path is increased because the permeability of air in the notch is very much lower than the permeability of the harmonic damper. This relatively high reluctance through the notch causes the magnetic flux to decrease and produces a change in V_0.

**Figure 5-16.
Crankshaft Position
Sensor**

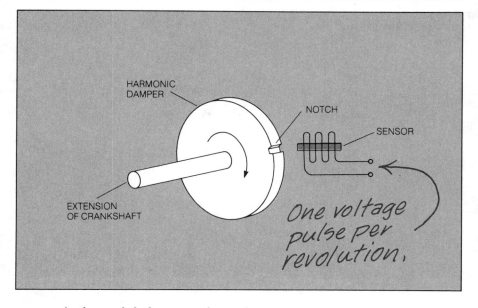

HARMONIC
DAMPER

NOTCH

SENSOR

EXTENSION
OF CRANKSHAFT

*One voltage
pulse per
revolution.*

As the crankshaft rotates, the notch passes under the sensor once
each engine revolution. The magnetic flux abruptly decreases, then increases
as the notch passes the sensor. This generates a voltage pulse which can be
used on electronic control systems to set ignition timing.

Hall-effect Position Sensor

The Hall-effect position
sensor also uses magnets
and a steel disk with tabs
to sense crankshaft posi-
tion. It usually is located
in the distributor and actually senses camshaft
position.

As mentioned above, one of the main disadvantages of the **magnetic**
reluctance sensor is its lack of output when the engine isn't running. A
crankshaft position sensor which avoids this problem is the Hall-effect position
sensor. This sensor is normally located in the distributor where it measures
camshaft position rather than crankshaft position. This sensor is relatively
inexpensive and requires only small modifications to a conventional distributor
with ignition points to use it. In fact, with appropriate circuitry, this sensor
also can be used to replace distributor ignition points.

A Hall-effect position sensor is shown in *Figure 5-17*. This sensor
is similar to the reluctance sensor in that it employs a steel disk having
protruding tabs and a magnet for coupling the disk to the sensing element.
Another similarity is that the steel disk varies the reluctance of the magnetic
path as the tabs pass between the magnet pole pieces.

**Figure 5-17.
Hall-Effect Position
Sensor**

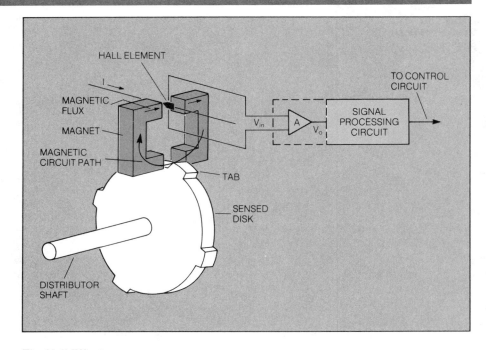

The Hall Effect

The Hall element is a small, thin, flat slab of semiconductor material. When a current, I, is passed through this slab by means of an external circuit as shown in *Figure 5-18a*, a voltage is developed across the slab perpendicular to the direction of current flow and perpendicular to the direction of magnetic flux. This voltage is proportional to both the current and magnetic flux density which flows through the slab. This effect—the generation of a voltage that is dependent on a magnetic field—is called the Hall effect.

In *Figure 5-18b*, the current, I, is represented by electrons, e, which have negative charge, flowing from left to right. The magnetic field, B, is perpendicular to the page and into the page. This is indicated by the arrows into the page of *Figure 5-18b*. Whenever an electron moves through a magnetic field, a force (called the Lorentz force) is exerted on the electron which is proportional to the electron velocity and the strength of the magnetic flux. The direction of this force is perpendicular to the direction of the magnetic flux lines and perpendicular to the direction in which the electron is moving. In *Figure 5-18b*, the Lorentz force direction is such that the electrons are deflected toward the lower sense electrode. Thus, this electrode is more negative than the upper electrode and a voltage exists between the electrodes having the polarity shown in *Figure 5-18b*.

The Hall element is a thin slab of semiconductor material which is placed between the magnets so it can sense the magnetic flux variations as the tab passes. A constant current is passed through the semiconductor in one direction and a voltage is generated that varies with the strength of the magnetic flux.

**Figure 5-18.
The Hall Effect**

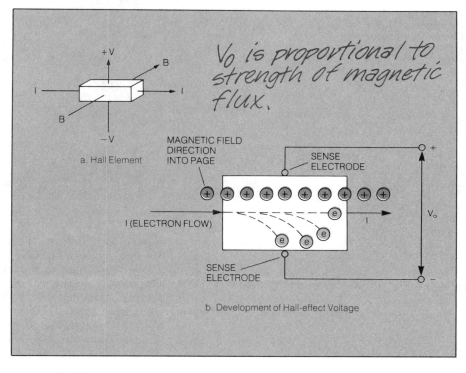

a. Hall Element

V_o is proportional to strength of magnetic flux.

MAGNETIC FIELD
DIRECTION
INTO PAGE

SENSE
ELECTRODE

I (ELECTRON FLOW)

SENSE
ELECTRODE

V_o

b. Development of Hall-effect Voltage

As the strength of the magnetic flux density increases, more of the electrons are deflected downward. If the current, I, is held constant, then the voltage, V_o, is proportional to the strength of the magnetic flux density. This voltage tends to be relatively weak so it is amplified as shown in *Figure 5-17*.

Output Waveform

It was shown in the discussion of the reluctance crankshaft position sensor that the magnetic flux density for this configuration depends upon the position of the tab. Recall that the magnetic flux is largest when one of the tabs is positioned symmetrically between the magnet pole pieces and that this position normally corresponds closely to TDC of one of the cylinders.

The voltage, V_o, waveform which is produced by the Hall element in the position sensor of *Figure 5-17* is illustrated in *Figure 5-19*. Since V_o is proportional to the magnetic flux density, it reaches maximum when any of the tabs is symmetrically located between the magnet pole pieces (i.e. corresponding to TDC of a cylinder). If the disk is driven by the camshaft, then the disk must have as many tabs as the engine has cylinders. Therefore, the disk shown would be for a 4-cylinder engine. It is important to realize that voltage output vs crankshaft angle is independent of engine speed. Thus, this sensor can be used for setting the engine timing when the engine is not running.

Because the Hall-effect sensor produces the same output voltage waveform regardless of engine speed, the engine timing can be set when the engine is not running.

**Figure 5-19.
Waveform of Hall-
Element Output Voltage
for Position Sensor of
Figure 5-17**

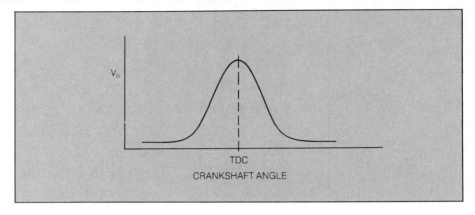

Shielded Field Sensor

Figure 5-20 shows another concept that uses the Hall-effect element in a way different from that just discussed. In this method, the Hall element is normally exposed to a magnetic field and produces an output voltage. When one of the tabs passes between the magnet and the sensor element, the low reluctance of the tab and disk provides a path for the magnetic flux which bypasses the Hall-effect sensor element. Since the sensor element is shielded from the magnetic field, the Hall-effect voltage and sensor output drops to near zero. Note in *Figure 5-20b* that the waveform is just opposite of the one in *Figure 5-19*.

**Figure 5-20.
Hall-Effect Position
Sensor that Shields the
Magnetic Circuit**

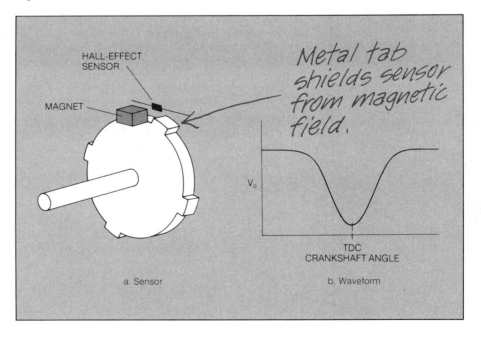

Optical Crankshaft Position Sensor

In the optical crankshaft position sensor, a disk coupled to the crankshaft has holes to pass light between the LED and the phototransistor. An output pulse is generated as each hole passes.

Shaft position can also be sensed using optical techniques. *Figure 5-21* illustrates such a system. Again, as with the magnetic system, a disk is directly coupled to the crankshaft. This time, the disk has holes in it that correspond to the number of tabs on the disks of the magnetic systems. Mounted on each side of the disk are fiber-optic light pipes. The hole in the disk allows transmission of light through the light pipes from the light-emitting diode source to the phototransistor used as a light sensor. Light would not be transmitted from source to sensor when there is no hole. The solid disk blocks the light. As shown in *Figure 5-21*, the pulse of light is detected by the phototransistor and coupled to an amplifier to obtain a satisfactory signal level. The output pulse level can very easily be standard TTL logic levels of +2.4V for the high level and +0.2V for the low level. Used as pulses, the signals provide time referenced pulses that can be signal processed easily with digital integrated circuits.

**Figure 5-21.
Optical Position Sensor**

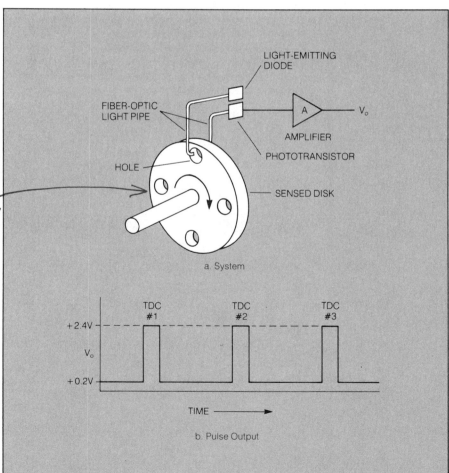

Rotation of disk causes alternate blocking and transmission of light.

a. System

b. Pulse Output

One of the problems with optical sensors is that they must be protected from dirt and oil; otherwise, they will not work properly. They have the advantage that they can sense position without the engine running and the pulse amplitude is constant with variation in speed.

AIR FLOW RATE SENSOR

To obtain the correct ratio of air to fuel, an important quantity which must be measured is the mass flow rate of air into the engine. This is especially true for a fuel control system that is operating in the open loop mode. Some sensors do not measure mass flow rate directly, but measure the velocity of the air through a given volume and then, using a table for the density of air at a given temperature, the mass flow rate of air is computed.

However, there are a variety of sensor configurations which measure volume flow rate directly. One of these, shown in *Figure 5-22*, uses a spring-loaded movable sensor flap which is mounted in the intake system. Air entering the intake exerts a force on the movable flap, but this force is opposed by the force produced by a spring connected to the flap. The deflection of the flap increases as the volume flow rate of the air increases.

The amount of deflection of the flap is measured by an electrical circuit which includes a potentiometer whose wiper control is physically coupled to the flap. The wiper arm of the potentiometer rotates about a common axis as it moves with the air flow sensor flap.

One kind of air flow rate sensor uses a spring-loaded movable flap. As the air flow moves the flap against the spring force, a mechanical linkage causes the potentiometer arm to wipe across the resistance element.

**Figure 5-22.
Cross Section of
Simplified Air-Flow
Sensor**

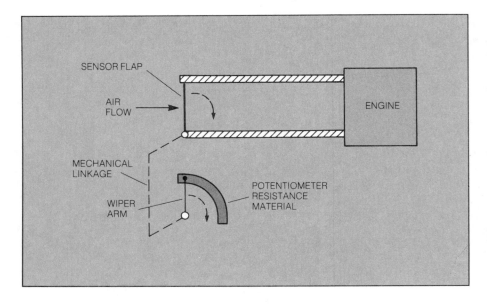

The potentiometer becomes a voltage divider when a voltage is applied across its end terminals. The output voltage varies between zero volts at no air flow to maximum voltage at maximum air flow.

The operation of the potentiometer can be understood with reference to *Figure 5-23*. The potentiometer consists of resistance material (e.g. resistance wire) on an insulating base which contacts a movable wiper arm (*Figure 5-23a*). The potentiometer acts as a voltage divider for the voltage applied between the end terminals of the resistance material. The sensor output voltage is taken from between the wiper arm and the end terminal connected to electrical ground (*Figure 5-23b*). The wiper voltage varies between zero volts for no air flow to the reference voltage when the flap is at maximum deflection. At intermediate angles, the wiper voltage is proportional to the flap deflection angle, which is dependent on the air flow rate. Therefore, the sensor output voltage can be used to indicate the air-flow rate.

**Figure 5-23.
Potentiometer
Configuration for Air-
Flow Sensor**

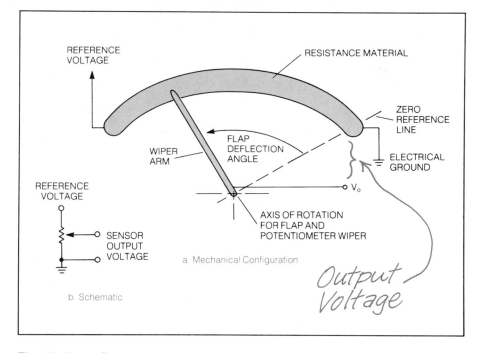

a. Mechanical Configuration

b. Schematic

Throttle Angle Sensor

A throttle angle sensor can use a design similar to the air-flow sensor. The throttle plate is mechanically linked to a potentiometer wiper arm and the output voltage represents the angle of the throttle plate.

Still another variable which must be measured for an electronic fuel control application is throttle angle or the position of the throttle plate. The operation of a commonly used throttle angle sensor is much like the airflow sensor described above. The wiper arm of a potentiometer is connected directly to the throttle shaft and rotates with it. A reference voltage is applied across the potentiometer; thus, the wiper voltage, which is the throttle position sensor output voltage, indicates the throttle angle. Throttle angle determines the amount of air entering the system for mixing with the fuel.

TEMPERATURE SENSORS

Temperature is an important parameter throughout the automotive system. The operation of an electronic fuel control system requires one to know the temperature of the coolant, the temperature of the inlet air and the temperature of the exhaust gas oxygen sensor, a sensor to be discussed a bit later. Several sensor configurations are available for measuring these temperatures, but we can illustrate the basic operation of most of the temperature sensors by explaining the operation of a typical coolant sensor.

One kind of coolant sensor uses a temperature-sensitive semiconductor called a thermistor. The sensor is typically connected as a varying resistance across a fixed reference voltage. As the temperature increases, the output voltage decreases.

Typical Coolant Sensor

A typical coolant sensor, shown in *Figure 5-24*, consists of a thermistor mounted in a housing which is designed to be inserted in the coolant stream. This housing is typically threaded with pipe threads which seal the assembly against coolant leakage.

**Figure 5-24.
Coolant Temperature
Sensor**

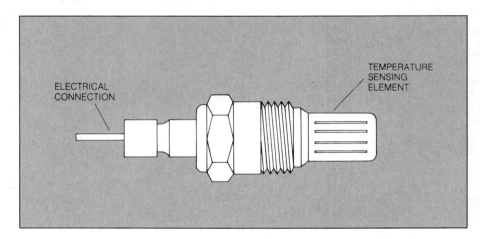

A thermistor is made of semiconductor material whose resistance varies inversely with temperature. For example at –40°C, a typical coolant sensor has a resistance of 100,000 ohms. The resistance decreases to about 70,000 ohms at 130°C.

The sensor is typically connected in an electrical circuit like that shown in *Figure 5-25* where the coolant temperature sensor resistance is denoted R_T. This resistance is connected to a reference voltage through a fixed resistance R. The sensor output voltage, V_T, is given by:

$$V_T = \left(\frac{R_T}{R + R_T} \right) V$$

The sensor output voltage varies inversely with temperature; that is, the output voltage decreases as temperature increases.

**Figure 5-25.
Typical Coolant
Temperature Sensor
Circuit**

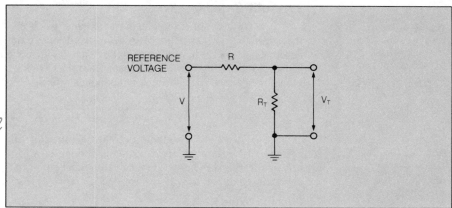

*V_T decreases
as temperature
sensed by
R_T increases.*

EXHAUST GAS EMISSION

It was noted at the beginning of this chapter that air and fuel are the basic inputs to an internal combustion automotive engine and that the outputs are the drive power and the exhaust emissions. It is very important to control the quantity of these exhaust emissions. Most electronic control systems use a reference point called stoichiometry in the scheme to control exhaust emissions.

Stoichiometry

Stoichiometry is the ideal air/fuel ratio at which the amount of air and the amount of fuel are a perfect match for perfect combustion. In theory, this ratio is 14.7:1.

The *air/fuel ratio* at stoichiometry is that mixture of air and fuel *in which, when ignited, all of the carbon and hydrogen would completely burn* yielding only carbon dioxide and water in the exhaust if combustion were perfect. The theoretical ratio of air mass to fuel mass for which this would occur is 14.7:1; that is, 14.7 pounds of dry air for 1 pound of gasoline.

An air/fuel ratio which exceeds stoichiometry means that more air is present than is needed for perfect combustion. In this situation, oxygen would be left in the exhaust because not all of the intake oxygen would be combined with carbon or hydrogen. A mixture with this air/fuel ratio is called a lean mixture.

Conversely, when the air/fuel ratio is less than stoichiometry, the amount of oxygen is insufficient to react with all of the carbon and hydrogen in the fuel. In this case, the exhaust gas contains unburned hydrocarbons (unburned fuel). A mixture with this air/fuel ratio is called a rich mixture.

EXHAUST GAS OXYGEN SENSOR

An exhaust gas oxygen sensor indirectly measures the air/fuel ratio by sensing the amount of oxygen in the exhaust gas.

The amount of oxygen in the exhaust gas is used as an indirect measurement of the air/fuel ratio. As a result, one of the most significant automotive sensors in use today is the exhaust gas oxygen (EGO) sensor. This sensor is often called a lambda sensor from the Greek letter lambda (λ), which is commonly used to denote the equivalence ratio:

$$\lambda = \frac{air/fuel}{air/fuel\ at\ stoichiometry}$$

When the air-fuel mixture has too much air, the condition is represented by lambda greater than one (denoted $\lambda > 1$). Conversely, when the air-fuel mixture has too little air (too much fuel), the condition is represented by an equivalence ratio of lambda less than one ($\lambda < 1$).

Zirconia Oxide EGO Sensor

The two types of EGO sensors in use today are based upon the use of active oxides of materials. One uses zirconia oxide (ZrO_2) and the other uses titanium oxide (TiO_2). *Figure 5-26* is a photograph of a typical ZrO_2 EGO sensor and *Figure 5-27* shows the physical structure. *Figure 5-27* indicates that a voltage, V_o, is generated across the ZrO_2 material. This voltage depends upon the engine air/fuel ratio.

**Figure 5-26.
Zirconia Oxide (ZrO_2)
EGO Sensor**

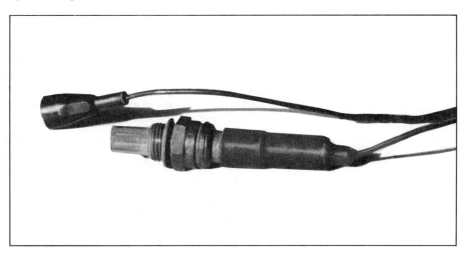

**Figure 5-27.
EGO Mounting and
Structure**

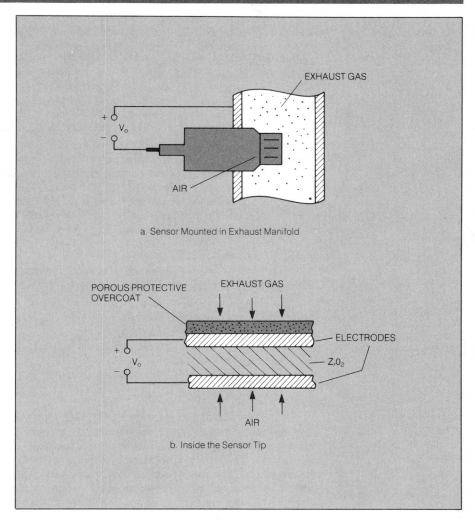

a. Sensor Mounted in Exhaust Manifold

b. Inside the Sensor Tip

The zirconia oxide EGO sensor uses zirconia oxide sandwiched between two platinum electrodes. One electrode is exposed to exhaust gas and the other is exposed to normal air for reference.

In essence, the EGO sensor consists of a thimble-shaped section of ZrO_2 with thin platinum electrodes on the inside and outside of the ZrO_2. The inside electrode is exposed to air and the outside electrode is exposed to exhaust gas through a porous protective overcoat.

A simplified explanation of EGO sensor operation is based upon the distribution of oxygen ions. An ion is an electrically charged atom. Oxygen ions have two excess electrons and each electron has a negative charge; thus, oxygen ions are negatively charged. The ZrO_2 has a tendency to attract the oxygen ions and they accumulate on the ZrO_2 surface just inside the platinum electrodes.

Because the exhaust contains fewer oxygen ions than air, the "air" electrode becomes negative with respect to the "exhaust" electrode. The voltage developed across the electrodes depends on the number of oxygen ions present in the air/fuel mixture.

The platinum plate on the air reference side of the ZrO_2 is exposed to a much higher concentration of oxygen ions than the exhaust gas side. The air reference side becomes electrically more negative than the exhaust gas side; therefore, an electric field exists across the ZrO_2 material and a voltage, V_o, results. The polarity of this voltage is positive on the exhaust gas side and negative on the air reference side of the ZrO_2. The magnitude of this voltage depends upon the concentration of oxygen in the exhaust gas and upon the sensor temperature.

The quantity of oxygen in the exhaust gas is represented by the oxygen partial pressure. Basically, this partial pressure is that proportion of the total exhaust gas pressure (nearly at atmospheric pressure) which is due to the quantity of oxygen. The exhaust gas oxygen partial pressure for a rich mixture varies over the range of 10^{-16} to 10^{-32} of atmospheric pressure. The oxygen partial pressure for a lean mixture is roughly 10^{-2} atmosphere.

If one wishes a more detailed discussion of this very complex reaction, consult the references given below.[1,2]

Desirable EGO Characteristics

An *ideal* EGO sensor would have an abrupt, rapid, and significant change in output voltage as the mixture passes through stoichiometry. The output voltage would not change as exhaust gas temperature changes.

The EGO sensor characteristics which are desirable for the type of limit cycle fuel control system we will discuss in Chapter 6 are:

1. Abrupt change in voltage at stoichiometry,
2. Rapid switching of output voltage in response to exhaust gas oxygen changes,
3. Large difference in sensor output voltage between rich and lean mixture conditions, and
4. Stable voltages with respect to exhaust temperature.

Switching Characteristics

Hysteresis is the difference in the switching point of the output voltage with respect to stoichiometry as a mixture passes from lean to rich, as contrasted to a mixture that passes from rich to lean.

The switching time for the EGO sensor also must be considered in control applications. An ideal characteristic for a limit cycle controller is shown in *Figure 5-28*. The actual characteristics of a new EGO sensor are shown in *Figure 5-29*. This data was obtained by slowly varying air/fuel across stoichiometry. The arrow pointing down indicates the change in V_o as air/fuel was varied from rich to lean. The up-arrow indicates the change in V_o as air/fuel ratio was varied from lean to rich. Note that the sensor output doesn't change at exactly the same point for increasing air/fuel ratio as for decreasing air/fuel ratio. This phenomenon is called hysteresis.

[1]Young, C.T., Bode, J.D., *Characteristics of $ZrO_{2\text{-}type}$ Oxygen Sensors for Automotive Applications*, SAE Publ. 790143, 1979.

[2]Young, C.T., *Experimental Analysis of ZrO_2 Oxygen Sensor Transient Switching Behavior*, SAE paper 810380, 1980.

Figure 5-28.
Ideal EGO Switching
Characteristics

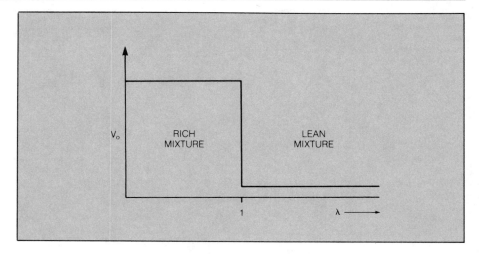

Figure 5-29.
Typical EGO Sensor
Characteristics

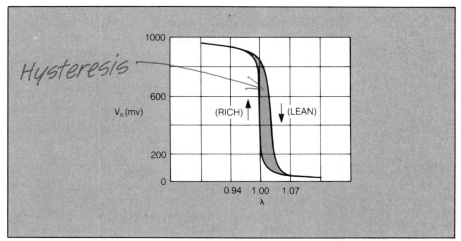

Temperature affects switching times and output voltage. Switching times at two temperatures are shown in *Figure 5-30*. (Note that the time per division is twice as much for the display at 350°C.) At an exhaust temperature of 350°C, the switching times are roughly 0.1 second whereas at 800°C they are about 0.05 second. This is a 2:1 change in switching times due to changing temperature.

Figure 5-30.
Typical Voltage
Switching
Characteristics of EGO

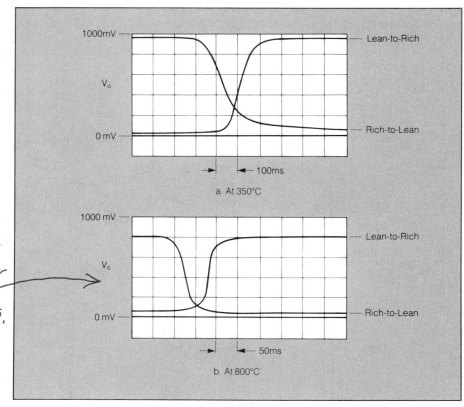

a. At 350°C

b. At 800°C

EGO sensors switch faster at higher temperatures.

The temperature dependence of the EGO sensor output voltage is very important. The graph in *Figure 5-31* shows the temperature dependence of an EGO sensor output voltage for lean and rich mixtures and for two different load resistances; 5 Megohms (million ohms) and 0.83 Megohms. The EGO sensor output voltage for a rich mixture is in the range from about 0.80 to 1.0 volt for an exhaust temperature range of 350 to 800°C. For a lean mixture, this voltage is roughly in the range of 0.05 to 0.07 volt for the same temperature range.

In Chapter 6, under certain conditions, the fuel control system using an EGO sensor will be operated open loop and for other conditions it will be operated closed loop. The EGO sensor should not be used for control at temperatures below about 300°C, because the difference between rich and lean voltages decreases rapidly with temperature in this region. This important property of the sensor is partly responsible for the requirement to operate the fuel control system in the open loop mode at low exhaust temperature. Closed loop operation with the EGO output voltage used as the error input cannot begin until the EGO sensor temperature exceeds about 300°C.

EGO sensors are not used for control when exhaust gas temperature falls below 300°C because the voltage difference between rich and lean conditions are minimal in this range.

**Figure 5-31.
Typical Influence of
Mixture and Temperature
on EGO Output Voltage**

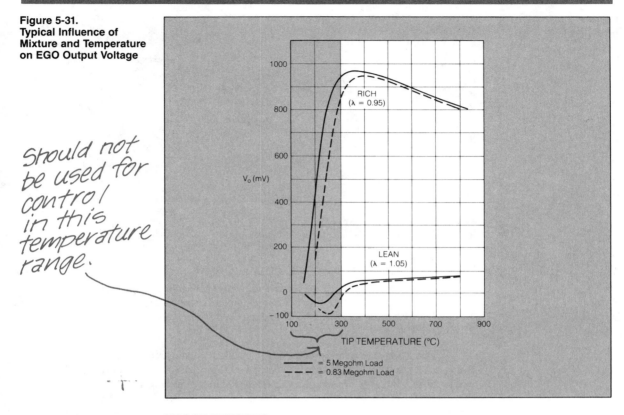

Should not be used for control in this temperature range.

KNOCK SENSORS

For certain electronic engine control systems, it is desirable to have a sensor which can detect an engine occurrence called "knock". Although a detailed discussion of knock is beyond the scope of this book, it can be described generally as a rapid rise in cylinder pressure during combustion. It does not occur normally, but under special conditions. It occurs most commonly with high manifold pressure and excessive spark advance. It is important to detect knock and avoid excessive knock; otherwise, there will be damage to the engine.

Some engine knock sensors use rods within a magnetic field to detect the presence of knock. Others use vibration-sensitive crystals or semiconductors.

One way of controlling the knocking is to sense when knocking begins and then retard the ignition until the knocking stops. A key to the control loop for this is a knock sensor. A knock sensor using magnetostrictive techniques is shown in *Figure 5-32*. When sensing knock, the magnetostrictive rods, which are in a magnetic field, change the flux field in the coil. This change in flux produces a voltage change in the coil. Other sensors use piezoelectric crystals, or the piezoresistance of a doped silicon semiconductor. Whichever type is used, it forms a closed loop system that retards the ignition to reduce the knock detected at the cylinders. Systems using knock sensors have been

Figure 5-32.
Knock Sensor

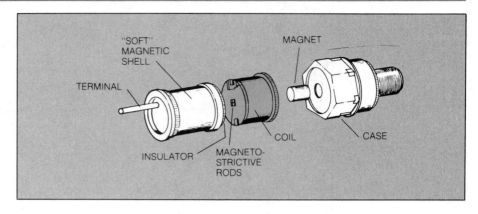

experimental to date and are still under development. The problem of
detecting knock is complicated by the presence of other vibrations and noises
in the engine.

AUTOMOTIVE ENGINE CONTROL ACTUATORS

In addition to the set of sensors, the electronic engine control is
critically dependent upon a set of actuators to control air/fuel ratio, ignition
and EGR. Each of these devices will be discussed separately.

FUEL METERING ACTUATOR

In engine control systems, it is necessary to maintain the air/fuel
ratio at or very near 14.7:1. This is accomplished by controlling fuel flow with a
fuel metering actuator. The two major classes of fuel metering actuators are
the electronic carburetor and the throttle body fuel injector (TBFI). The
electronic carburetor has been in production for some time, but it will probably
be replaced in the future by the TBFI.

Electronic Carburetor

In an electronic car-
buretor, the metering rod
which adjusts the fuel flow
is controlled by a vacuum
regulator which receives
its control signals from
the electronic engine
controller.

The electronic carburetor is similar to a conventional carburetor
except that the fuel metering rods can be electrically controlled. *Figure 5-33*
is a drawing of a typical electronic carburetor. The mechanism for varying
air/fuel ratio in response to an electrical command signal is provided by an
electromechanical vacuum regulator and a special set of metering rods whose
position is controlled by vacuum.

The vacuum regulator is an actuator which supplies a control vacuum
signal ranging from zero to about 8 inches of mercury depending upon the
command from the electronic controller. The vacuum regulator is designed
so that the control vacuum varies inversely with the command current. This
control vacuum remains constant for a given command current provided the
source vacuum (manifold vacuum) exceeds the control vacuum by about one
inch of mercury. *Figure 5-34* is a graph of the control vacuum versus current
for a typical vacuum regulator.

**Figure 5-33.
Simplified Electronic
Carburetor**

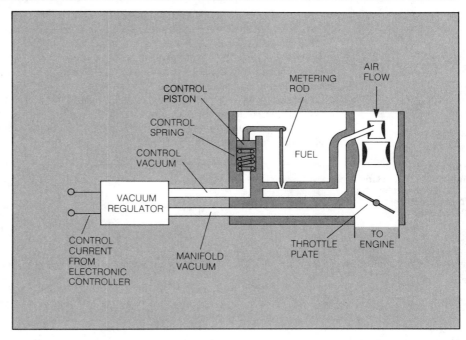

**Figure 5-34.
Typical Vacuum
Regulator Control
Characteristic Curve**

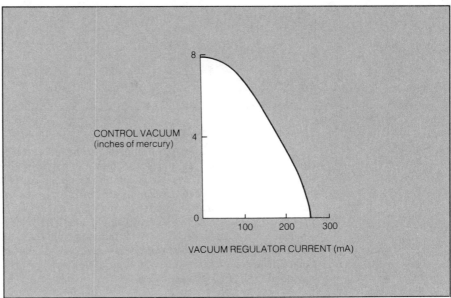

With no control vacuum, the metering rod is held open by a spring, resulting in maximum fuel flow. As control vacuum increases, the metering rod closes to reduce fuel flow.

The control piston and spring in the carburetor move the metering rods between their extreme limits. The spring holds the metering rods open with no vacuum present. When control vacuum is applied, it pulls the piston against the opposing spring pressure. As the piston moves down, it pulls the metering rod down to allow less fuel to pass. As the control vacuum increases, the metering rod closes more to allow a lower rate of fuel flow into the moving air stream. *Figure 5-35* is a graph of air/fuel ratio versus control vacuum for a typical electronic carburetor. Notice that the air/fuel ratio can be varied electrically from just under 14 to about 17. This is a sufficient range for electronic fuel control applications.

**Figure 5-35.
Typical Air/Fuel Ratio
Control Characteristic
for an Electronic
Carburetor**

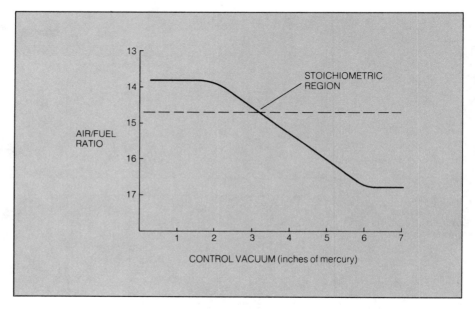

Throttle Body Fuel Injection

A throttle body fuel injector replaces the carburetor. It uses injectors to spray pulses of pressurized fuel into the moving air stream that enters the intake manifold rather than letting the fuel be continually drawn from a bowl by manifold vacuum.

The other major type of fuel metering actuator is the throttle body fuel injector (TBFI). This consists of a housing having one or more bores through which the air flows into the intake manifold. A pulsed electric fuel injector is mounted in each bore to inject fuel into the moving air stream. The throttle plate and throttle position sensor also are part of this assembly.

Figure 5-36 shows a typical single bore TBFI and *Figure 5-37* is a schematic drawing of a TBFI. Fuel is pumped from the tank and enters the TBFI unregulated, but the pressure is regulated to about 10 pounds per square inch by a fuel pressure regulator assembly. The pressure regulated fuel is applied to the fuel injector.

**Figure 5-36.
Single Bore TBFI**

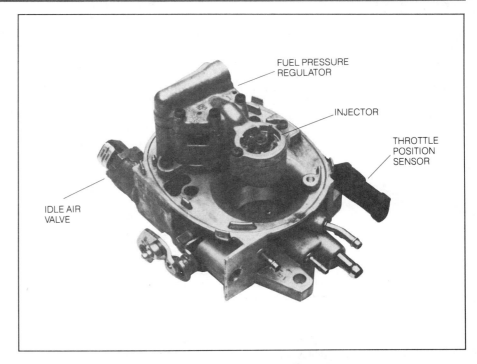

FUEL PRESSURE
REGULATOR

INJECTOR

THROTTLE
POSITION
SENSOR

IDLE AIR
VALVE

**Figure 5-37.
Throttle Body Fuel
Injection (TBFI) Fuel
Metering Schematic**

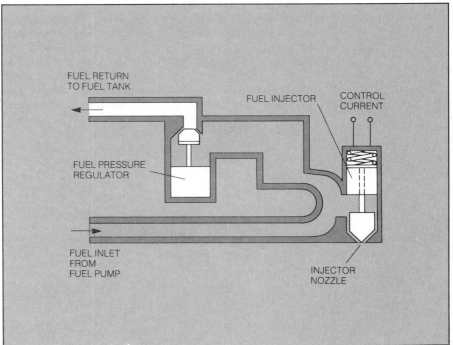

FUEL RETURN
TO FUEL TANK

FUEL INJECTOR

CONTROL
CURRENT

FUEL PRESSURE
REGULATOR

FUEL INLET
FROM
FUEL PUMP

INJECTOR
NOZZLE

The fuel injector consists of a spray nozzle and a solenoid operated plunger. Whenever the plunger is against the nozzle, fuel is prevented from flowing. Whenever the plunger is lifted from the nozzle, fuel flows at a fixed rate through the nozzle into the air stream going to the intake manifold. Thus, the plunger acts as a fuel injection on-off valve.

A throttle body fuel injector contains a solenoid, a spring-loaded plunger, and an injector nozzle that act as an on-off valve under electrical control.

The plunger position is controlled by a solenoid and a spring. The plunger is held down tightly against the nozzle by a powerful spring when no current is applied to the solenoid. Whenever a current of sufficient magnitude is applied to the control current terminals, a magnetic field is produced in the solenoid which draws the plunger away from the nozzle. This takes a certain amount of time and this becomes important, as will be seen later. However, for now, let's consider that the plunger motion is so rapid that the change in plunger position can be considered instantaneous. Hence, the solenoid, plunger and nozzle act as an electrically switched valve. This valve is either closed or open depending upon whether the control current is off or on, respectively.

The fuel flow rate through the nozzle is constant for a given regulated fuel pressure and nozzle geometry; therefore, the quantity of fuel injected into the air stream is proportional to the time the valve is open. The control current which operates the fuel injector is pulsed on and off and the air/fuel ratio is proportional to the duty cycle of the pulse train from the electronic controller.

Fuel-Injector Signal

For a pulse train signal, the ratio of on time, t, to the period of the pulse, T (on time plus off time), is called the duty cycle. This is shown in *Figure 5-38*. The fuel metering actuator is energized for time t to allow fuel to spray from the nozzle into the air stream going to the intake manifold. The actuator is de-energized for the remainder of the period. Therefore, a low duty cycle, as shown in *Figure 5-38a*, is used for a high air/fuel ratio (lean mixture) and a high duty cycle as shown in *Figure 5-38b* for a low air/fuel ratio (rich mixture).

During the open loop mode of operation, a fixed duty cycle is chosen by the electronic controller to supply the basic air/fuel ratio. When the system switches to the closed loop mode, the electronic controller varies the pulse width (t) to vary the duty cycle to vary the air/fuel ratio in response to the error signal from the EGO sensor.

Figure 5-38.
Pulse Mode Fuel Control
Signal to Fuel Metering
Actuator

High air-to-fuel ratio.

Low air-to-fuel ratio.

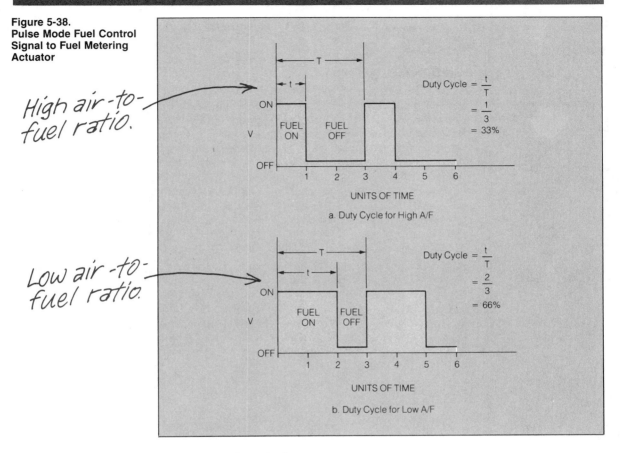

a. Duty Cycle for High A/F

b. Duty Cycle for Low A/F

Low and High Duty Cycle

Because the flow of fuel through the nozzle is either on or off, the air/fuel ratio must be regulated by controlling the duty cycle; that is, the ratio of injector on time to total on-and-off time.

The frequency of the pulses (pulses per second) of the duty cycle signal for the actuator is determined by engine RPM. As engine speed increases, the pulse frequency increases. However, at some upper frequency limit which is determined by the duty cycle, the controller holds the frequency at a fixed value. This is because the period (T) of the pulse signal takes less time as frequency increases; thus, the on time (t) also gets shorter in real time even if the duty cycle remains constant. It takes a certain amount of time for the solenoid valve to open and a certain amount of time for it to close. Therefore, as t gets shorter with increasing frequency and a low duty cycle, a point is reached where the on time is not long enough to open the valve. Similarly, for high frequency and a high duty cycle, the off time is not sufficient for the valve to close. If the frequency were not limited, the operation of the valve could become erratic and unreliable.

The duty cycle signal can be sent to the TBFI actuator directly from the computer or by a special purpose interface which converts a binary number into a duty cycle output.

IGNITION ACTUATOR

Strictly speaking, the actuator in an ignition system should be viewed as a coil, distributor, and spark plugs in combination. Interrupting the coil primary current produces an action—ignition of the air and fuel mixture. However, many electronic ignition systems package the coil and associated electronics in a single module. With this in mind, the term actuator in ignition systems is expanded to include the associated electronic circuitry as part of the actuator.

With this viewpoint, *Figure 5-39* illustrates an electronic ignition circuit. Chapter 6 will have further discussion of ignition systems and the control of spark ignition and timing; however, a typical actual actuator circuit is shown in *Figure 5-40*. Note that the actuator receives its control pulse from an ignition timing sensor of the type discussed earlier.

The combination of components that make up a complete ignition system can be viewed overall as an actuator; that is, the input electrical signals produce an action (the ignition of the air/fuel mixture).

**Figure 5-39.
Ignition Actuator**

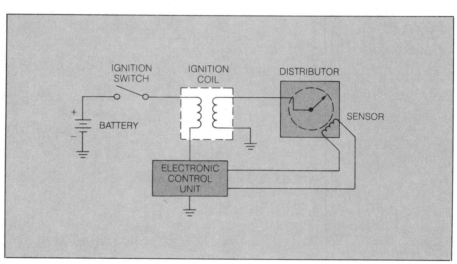

EXHAUST GAS RECIRCULATOR ACTUATOR

Another important actuator for modern emission controlled engines is the exhaust gas recirculator (EGR) actuator. EGR refers to the addition of exhaust gas to the intake charge to lower NO_x emissions. The EGR actuator is a valve which essentially connects the intake and exhaust manifolds. When the valve is open, a portion of the exhaust gas flows into the intake manifold because of the lower pressure in the intake manifold.

Figure 5-40.
Electronic Control Unit

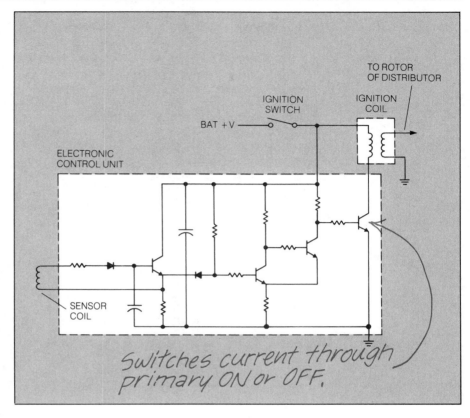

TO ROTOR
OF DISTRIBUTOR

IGNITION
SWITCH

IGNITION
COIL

BAT + V

ELECTRONIC
CONTROL UNIT

SENSOR
COIL

Switches current through primary ON or OFF.

Although there are many EGR configurations, only one representative example will be discussed to explain the basic operation of this type of actuator. The example EGR actuator is shown schematically in *Figure 5-41*. This actuator is a vacuum operated diaphragm valve with a spring that holds the valve closed if no vacuum is applied. The vacuum which operates the diaphragm is supplied by the intake manifold and is controlled by a solenoid operated valve. This solenoid valve is controlled by the output of the control system.

This solenoid operates essentially the same as that explained in the TBFI. Whenever the solenoid is energized (i.e. by current supplied by the control system flowing through the coil) the EGR valve is opened by the applied vacuum. Whenever the solenoid is de-energized, the vacuum is cut off from the EGR valve and the spring holds the EGR valve closed. The amount of EGR is controlled by the average time the solenoid is energized (i.e. duty cycle of the pulsed control current) as was described for the TBFI.

One kind of EGR actuator consists of a vacuum-operated valve with the vacuum supply controlled by a solenoid. When the EGR valve is open, exhaust gas flows into the intake manifold.

Figure 5-41.
EGR Actuator Control

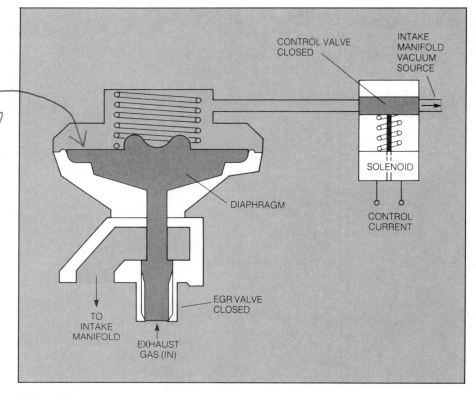

Vacuum pulls up diaphragm

CONTROL VALVE
CLOSED

INTAKE
MANIFOLD
VACUUM
SOURCE

SOLENOID

CONTROL
CURRENT

DIAPHRAGM

EGR VALVE
CLOSED

TO
INTAKE
MANIFOLD

EXHAUST
GAS (IN)

SUMMARY

In this chapter, we have examined some of the important automotive sensors and actuators for electronic control systems. We have attempted to explain the basic fundamentals of these sensors and to give some indication of how they are used. In addition, we have shown that existing sensors and actuators have certain performance limitations and often are more costly than desired.

Moreover, there is a fairly large set of physical variables for which no suitable sensor is presently available for on-board automobile application. Some of these are:

1) indicated or brake torque
2) exhaust gas concentration
3) air/fuel ratio
4) tire pressure

Various ideas for cost-effective sensors to measure these variables have been considered and tried, but none have yet been satisfactory. Indeed, there is opportunity for some inventive genius to develop a sensor for any one of these variables, or for significant improvement of existing sensors.

Quiz for Chapter 5

1. Which of the following operations is performed by a sensor?
   a. it selects transmission gear ratio
   b. it measures some variable
   c. it is an output device
   d. it sends signals to the driver

2. What does an actuator do?
   a. it is an input device for an engine control system
   b. it provides a mathematical model for an engine
   c. it causes an action to be performed in response to an electrical signal
   d. it indicates the results of a measurement

3. What is a MAP sensor?
   a. it is a sensor which measures manifold absolute pressure
   b. it is a vacation route planning scheme
   c. it is a measurement of fluctuations in manifold air
   d. it is an acronym for mean atmospheric pressure

4. What is an EGO sensor?
   a. it is a measure of the self-centeredness of the driver
   b. it is a device for measuring the oxygen concentration in the exhaust of an engine
   c. it is a spark advance mechanism
   d. it measures crankshaft acceleration

5. The crankshaft angular position sensor measures:
   a. the angle between the connecting rods and the crankshaft
   b. the angle between a line drawn through the crankshaft axis and a mark on the flywheel and a reference line
   c. the pitch angle of the crankshaft axis
   d. the oil pressure angle

6. The Hall effect is:
   a. the resonance of a long, narrow corridor
   b. the flow of air through the intake manifold
   c. zero crossing error in camshaft position measurements
   d. a phenomenon occurring in semiconductor materials in which a voltage is generated which is proportional to the strength of a magnetic field

7. A mass airflow sensor measures:
   a. the density of atmospheric air
   b. the composition of air
   c. the rate at which air is flowing into an engine measured in terms of its mass
   d. the flow of exhaust out of the engine

8. An electronic carburetor is:
   a. a carburetor having electronically controlled fuel metering rods
   b. a device for injecting electrons into the engine
   c. a carburetor having a mass airflow sensor
   d. a carburetor having an electronically controlled throttle

9. Throttle body fuel injection refers to:
   a. inserting fuel below the throttle plate
   b. a form of fuel metering actuator
   c. unregulated fuel flow
   d. a continuous flow fuel injection system

10. A thermistor is:
    a. a semiconductor temperature sensor
    b. a device for regulating engine temperature
    c. a temperature control system for the passenger compartment
    d. a new type of transistor

11. Intake manifold pressure:
    a. is constant
    b. fluctuates as air is drawn into each cylinder
    c. is produced by an external pump
    d. none of the above

12. An aneroid chamber is a:
    a. chamber holding high air pressure
    b. small tube connected to the intake manifold
    c. displacement sensor
    d. pair of diaphragms which are welded to form an evacuated chamber

13. Piezoresistivity is:
    a. a property of certain semiconductors in which resistivity varies with strain
    b. a resistance property of insulators
    c. metal bonding pads
    d. an Italian resistor

14. The resonant frequency, f_r, of a circuit is given by:
    a. $\dfrac{1}{2\pi\sqrt{LC}}$
    b. $2\pi\sqrt{LC}$
    c. $\dfrac{2\pi\sqrt{LC}}{c}$
    d. none of the above

15. Reluctance is:
    a. the reciprocal of permeability
    b. a property of a magnetic circuit which is analogous to resistance in an electrical circuit
    c. a line of constant magnetic flux
    d. none of the above

16. An optical crankshaft position sensor:
    a. senses crankshaft angular position
    b. operates by alternately passing or stopping a beam of light from a source to an optical detector
    c. operates in a pulsed mode
    d. all of the above

17. The resistance of a thermistor:
    a. varies inversely with temperature
    b. varies directly with temperature
    c. is always 100,000 ohms
    d. none of the above

18. Duty cycle in a fuel metering actuator refers to the ratio of:
    a. fuel ON time to fuel OFF time
    b. fuel OFF time to fuel ON time
    c. fuel ON time to fuel ON time plus fuel OFF time
    d. none of the above

19. An EGO sensor is:
    a. a perfectly linear sensor
    b. a sensor having two different output levels depending upon air/fuel ratio
    c. unaffected by exhaust oxygen levels
    d. unaffected by temperature

20. A potentiometer is:
    a. a variable resistance circuit component
    b. sometimes used to sense airflow
    c. can be used in a throttle angle sensor
    d. all of the above

The Basics of Electronic Engine Control

ABOUT THIS CHAPTER

In the preceding chapters, a foundation of fundamentals and basic concepts has been developed. It consisted of understanding the basic automotive functions and realizing where electronics might be used. It included basic system, electronic and computer fundamentals to help in the understanding of how they are used for automotive electronic control. Sensors and actuators were discussed to clarify how physical quantities are converted to electrical signals and electrical signals are converted to physical action.

With this knowledge in hand, the discussion shifts to applying this knowledge to understand how automotive functions are controlled electronically. We begin with the drive train, and in this chapter and the following chapter we will consider applying electronics to control the automotive engine.

Technically speaking, the majority of automotive engines are gasoline fueled, spark ignited, liquid cooled internal combustion engines. It was discussed in Chapter 1 and called an SI engine. The application of electronic technology to the control of such engines is a rather interesting story. Let's begin this discussion with a brief survey of the history of the engine and the developments which led to electronic control.

BRIEF HISTORY

Throughout most of the history of the automobile, the ignition system has been the only electrical engine function. Only recently have other engine functions been controlled electrically or electronically.

The gasoline fueled, spark ignited internal combustion engine which powers most automobiles came into being in Germany in 1876. It was the brainchild of Nicholaus Otto. It was modified and improved over the next 100 years, with the most significant improvements resulting from automotive needs, since the primary use of the Otto engine has been to power automobiles. In its entire history, from 1876 until the decade 1970-1980, the only electrically implemented primary engine function has been for ignition.

Suddenly in the 1970's, a variety of primary functions were being controlled electrically or electronically. It is natural to ask the question; "Why have electronic engine controls become so important to the automotive industry?" It is not because of a lack of confidence in the capability of electronic systems that they haven't been used previously. On the contrary, the capabilities and reliability of electronic systems have been spectacularly demonstrated by modern aircraft and spacecraft computers and communication systems.

Rather these systems haven't been used for automotive engine control because it has been possible for over 100 years to provide control of the engine using hydraulic or pneumatic controls (fluidic systems). In fact, by the decade 1950–1960, such controls had become quite sophisticated. Surely after many years of development and considerable cost, the automobile industry would be reluctant to give up a well-understood and well-defined technology to embark on the potentially risky program of implementing engine control electronically. Indeed, the answer to the question is both technical and economic and is worthy of a brief discussion.

MOTIVATION FOR ELECTRONIC ENGINE CONTROL

The motivation for electronic engine control came in part from two government requirements. The first came about as a result of legislation to regulate automobile exhaust emissions under the authority of the Environmental Protection Agency (EPA). The second was a thrust to improve the national average fuel economy by government regulation.

Exhaust Emissions

The engine exhaust consists of the products of combustion of the air and gasoline mixture. Gasoline is a mixture of chemical compounds which are termed hydrocarbons. This name is derived from the chemical formation of the various gasoline compounds, each of which is a chemical union of hydrogen (H) and carbon (C). In addition, the gasoline also contains natural impurities as well as chemicals added by the refiner. All of these can produce undesirable exhaust elements.

Following the perfect combustion of gasoline, exhaust gases contain only harmless carbon dioxide and water. However, after imperfect combustion in the typical engine, the exhaust contains harmful CO, NOx, unburned HC and other emissions.

During the combustion process, the carbon and hydrogen combine with oxygen from the air, releasing heat energy and forming various chemical compounds. If the combustion were perfect, then the exhaust gases would consist only of carbon dioxide, CO_2, and water, H_2O, neither of which are considered harmful in the atmosphere. In fact, these are present in a human's breath.

Unfortunately, the combustion of the SI engine is not perfect. In addition to the CO_2 and H_2O, the exhaust contains amounts of carbon monoxide, CO, oxides of nitrogen (chemical unions of nitrogen and oxygen which are denoted NO_x), unburned hydrocarbons (HC), oxides of sulfur and other compounds. The exhaust emissions controlled by government standards are CO, HC and NO_x.

Automotive exhaust emission control requirements started in the U.S. in 1966 when the California state regulations became effective. Since then, the federal government has imposed emission control limits for all states and the standards became progressively more difficult to meet through the decade 1970-1980. Auto manufacturers found that the traditional engine controls could not control the engine sufficiently to meet these emission limits while maintaining adequate engine performance, so they turned to electronic controls.

Fuel Economy

We all have some idea of what fuel economy means. It is related to the number of miles which we can drive our car for each gallon of gasoline consumed. We usually refer to it as miles per gallon (MPG) or simply "mileage". Just like it improves emission control, another important feature of electronic engine control is its ability to improve fuel economy.

It is well recognized by layman and expert alike that the mileage of a vehicle is not unique. It depends upon size, shape, weight, and how the car is driven. The best mileage is achieved under steady cruise conditions. City driving, with many starts and stops, yields worse mileage than steady highway driving. Of course, some combination of city driving and highway driving could represent a "typical" car trip.

The government fuel economy standards are not based on just one car, but are stated in terms of the average rated miles per gallon fuel mileage for the production of all models by a manufacturer for any year. This latter requirement is known in the automotive industry by the acronym CAFE (Corporate Average Fuel Economy). It is a somewhat complex requirement and is based upon measurements of the fuel used during a prescribed simulated standard driving cycle.

FEDERAL GOVERNMENT TEST PROCEDURES

For an understanding of both emission and CAFE requirements, it is helpful to review the standard cycle and how the emission and fuel economy measurements are made. The U.S. Federal Government has published test procedures which includes several steps. The automobile is first placed on a chassis dynamometer like the one shown in *Figure 6-1*.

A chassis dynamometer is a test stand that holds a vehicle. The vehicle might be a car, truck, etc. It is equipped with instruments capable of measuring the power which is delivered at the drive wheels of the vehicle under various conditions. The vehicle is held on the dynamometer so that it can't move when power is applied to the drive wheels. The drive wheels are in contact with two large rollers. One roller is mechanically coupled to an electric generator which has a means of varying the load on its electrical output. The other roller has instruments to measure and record the vehicle speed. The generator absorbs all mechanical power which is delivered at the drive wheels and the horsepower is calculated from the electrical output. 746 watts of electrical output equals one horsepower. The controls of the dynamometer can be set to simulate the correct load and inertia of the vehicle moving along a road under various conditions. The conditions are the same as if the vehicle actually were being driven.

**Figure 6-1.
Chassis Dynamometer**

Emission samples are collected and measured during a simulated urban trip containing a high percentage of stop-and-go driving.

The vehicle is operated according to a prescribed schedule of speed and load to simulate the specified trip. One is an urban trip and one is a highway trip. The schedules are shown in *Figure 6-2*. The 18 cycles of the urban simulated trip of *Figure 6-2a* take 1,372 seconds and include acceleration, deceleration, stops, starts and steady cruise, such as would be encountered in a 'typical' city automobile trip of 7.45 miles (12 km). The highway schedule of *Figure 6-2b* takes 765 seconds which simulates 10.24 miles (16.5 km) of highway driving.

During the operation of the vehicle in the urban test, the exhaust is continuously collected and sampled. At the end of the test, the absolute mass of each of the important exhaust gases is determined. The regulations are stated in terms of the total mass of each exhaust gas divided by the total distance of the simulated trip.

**Figure 6-2.
Federal Driving
Schedules**
(Title 40 United States Code of
Federal Regulations)

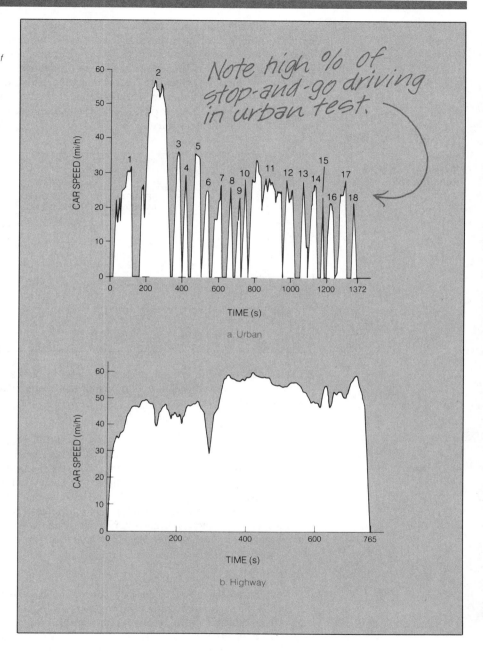

a. Urban

b. Highway

Fuel consumption also is measured during the tests. Emission and MPG requirements have grown increasingly stringent since 1968.

In addition to emission measurement, each manufacturer must determine the fuel consumption in miles per gallon for each type of vehicle and must compute the corporate average mileage for all vehicles of all types produced in a year. Fuel consumption is measured during both the urban and highway tests and the composite fuel economy is calculated.

Table 6-1 is a summary of the exhaust emission requirements and CAFE for representative years. It shows the emission reductions and increased fuel economy required.

**Table 6-1.
Emission and MPG
Requirement**

YEAR	FEDERAL HC/CO/NO$_x$	CALIFORNIA HC/CO/NO$_x$	CAFE MPG
1968	3.2/33/—	— — —	—
1971	2.2/23/—	— — —	—
1978	1.5/15/2.0	0.41/9.0/1.5	18
1979	1.5/15/2.0	0.41/9.0/1.5	19
1980	0.41/7.0/2.0	0.41/9.0/1.5	20

Because of these requirements, each manufacturer has a strong incentive to minimize exhaust emissions and maximize fuel economy for each vehicle produced.

Meeting the Requirements

Engines using mechanical, hydraulic, or pneumatic controls do not meet government regulations, but engines using electronic engine controls can.

Unfortunately, as we shall see later in the chapter, meeting the government regulations causes some sacrifice in performance. Moreover, attempts to meet the *Table 6-1* standards using mechanical, electromechanical, hydraulic or pneumatic controls like those used in the past have not been cost effective. In addition, such controls do not have the capability to reproduce functions with a good enough accuracy across a range of production vehicles, over all operating conditions, and over the life of the vehicle to stay within the tolerances required to meet the Environmental Protection Agency (EPA) regulations.

The Role of Electronics

This is where electronics takes over. The use of digital electronics control has the capability to meet the government regulations by controlling the system accurately with excellent tolerance. In addition, the system has long-term calibration stability. As an added advantage, this type of system is very flexible. Because it uses microcomputers, it can be modified through programming changes to meet a variety of different vehicle-engine combinations. Critical quantities that describe an engine can be changed easily by changing data that is stored in the system's computer memory.

Additional Cost Incentive

Dropping costs of micro-processors and other very large scale integrated circuits have made electronic engine control an increasingly attractive system for automobile manufacturers.

Besides providing the control accuracy and stability, there is a cost incentive to use digital electronics control. The system components, the multifunction digital integrated circuits, are decreasing in cost; thus, decreasing the system cost. During the decade of the 1970's, considerable investment has been made by the semiconductor industry for the development of low cost, multifunction integrated circuits. In particular, the microprocessor and microcomputer[1] have reached an advanced state of capability at relatively low cost. This has made the electronic digital control system for the engine as well as other on-board automobile electronic systems commercially feasible.

As pointed out in Chapter 3, the multifunction digital integrated circuits continue to be designed with more and more functional capability through very-large-scale integrated circuits (VLSI), the costs will continue to decrease and, at the same time, offer improved electronic system performance in the automobile.

In summary, the electronic engine control system duplicates the function of conventional fluidic control systems, but with greater precision, and has the capability of optimizing engine performance while meeting the exhaust emission and fuel economy regulations.

CONCEPT OF AN ELECTRONIC ENGINE CONTROL SYSTEM

An electronic engine control system is an assembly of electronic and electromechanical components which continuously vary the engine calibration (defined later) in order to satisfy government exhaust emission and fuel economy regulations. *Figure 6-3* is a block diagram of a generalized electronic engine control system.

**Figure 6-3.
General Electronic
Engine Control System**

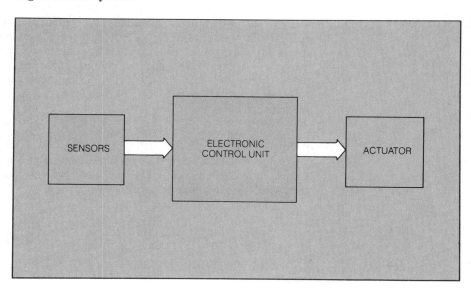

[1] Cannon, Luecke, *Understanding Microprocessors*, 1979, Texas Instruments Incorporated, Dallas, Texas.

As we explained in Chapter 2, a control system requires measurements of certain variables which tell the controller the state of the system being controlled. The electronic engine control system receives input electrical signals from the various sensors which measure the state of the engine. From these signals, the controller generates output electrical signals to the actuators which determine the engine calibration.

Examples of automotive engine control system sensors were discussed in Chapter 5. As mentioned, the configuration and control for an automotive engine control system is determined in part by the set of sensors which are available to measure the variables. In many cases, the sensors which are available for automotive use involve compromises between performance and cost. In other cases, only indirect measurements of certain variables are feasible.

Figure 6-4a identifies the automotive functions that surround the engine in a bit more detail. There is a fuel metering system to set the air-fuel mixture going into the engine through the intake manifold. As we mentioned before, when the air-fuel mixture is such that there is 14.7 times as much air by weight as fuel, the air/fuel ratio is at a reference point called stoichiometry. Spark control determines when the air-fuel mixture is ignited after it is compressed in the cylinders of the engine. Power drive occurs at the output drive shaft and the remains of combustion flow out the exhaust system.

In the exhaust system, there is a valve to control the amount of the exhaust gas being recirculated back to the input and a catalytic converter to further control emissions.

There is considerable variation between the configuration of electronic control systems used by various manufacturers over various model years. At one stage of development, the electronic engine control consisted of separate subsystems for fuel control, spark control, and exhaust gas recirculation. As shown in *Figure 6-4b*, it appears that the automotive engine control system is evolving toward an integrated digital system in which these subsystems are treated as separate functions of the same controller.

In this chapter, we will discuss the various electronic engine control functions separately and explain how each function is implemented by a separate control system. Then we'll show how they are being integrated into one system.

DEFINITION OF GENERAL TERMS

Before proceeding with the details on engine control systems, certain definitions of terms must be clarified. Let's take some general terms first and then some specific engine performance terms.

Overall engine functions that are subject to electronic engine controls are air/fuel ratio, spark control, and exhaust gas recirculation.

Earlier models of electronic control systems used separate controllers for each engine function. The current trend is towards a single, integrated controller that monitors multiple functions.

**Figure 6-4.
Engine Functions and
Control**

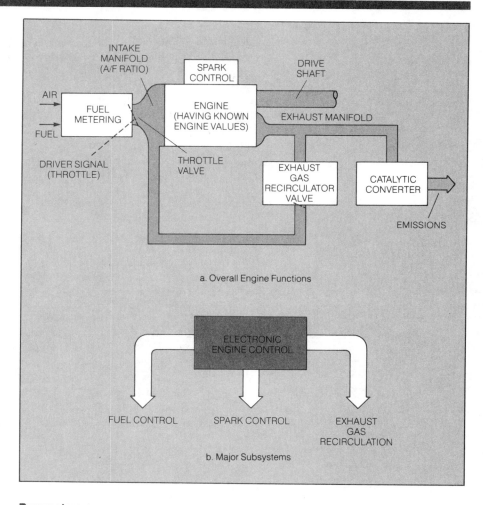

a. Overall Engine Functions

b. Major Subsystems

Parameters

Design parameters such as engine size, compression ratio, etc. are fixed; therefore, they are not subject to any engine operating control.

A parameter is a numerical value of some engine dimension that is fixed by design. Examples of engine design parameters include the piston diameter (bore), the distance the piston travels on one stroke (stroke), and the length of the crankshaft lever arm (throw). The bore and stroke determine the cylinder volume and the displacement. Displacement is the total volume of air which is displaced as the engine rotates through two complete revolutions. Compression ratio is the ratio of cylinder volume at BDC to the volume at TDC. Other parameters which engine designers must specify include combustion chamber shape, camshaft cam profile, intake and exhaust valve size, and valve timing. All of these design parameters are fixed by design and are not subject to control while the engine is operating.

Variables

A variable is a quantity which changes or may be changed as the engine operates. *Figures 6-5* and *6-6* identify specific engine control variables.

Inputs to Controllers

A variable is a quantity which can be changed as the engine operates.

Figure 6-5 identifies the major physical quantities that are control variables and are sensed and provided to the electronic controller as inputs. These include:

a. Throttle-valve position
b. Manifold Absolute Pressure (MAP)
c. Engine temperature (coolant temperature)
d. Engine speed (RPM)
e. Exhaust Gas Recirculation valve position
f. Oxygen content in exhaust gas.

**Figure 6-5.
Major Controller Inputs
from Engine**

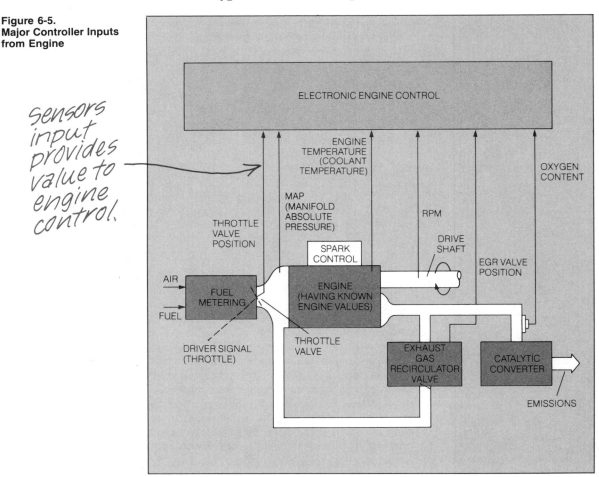

sensors input provides value to engine control.

Outputs from Controllers

Figure 6-6 identifies the major physical quantities that are outputs from the controller. These include:

a. Fuel metering control
b. Ignition control
c. Ignition timing
d. Exhaust Gas Recirculation control

**Figure 6-6.
Major Controller Outputs
to Engine**

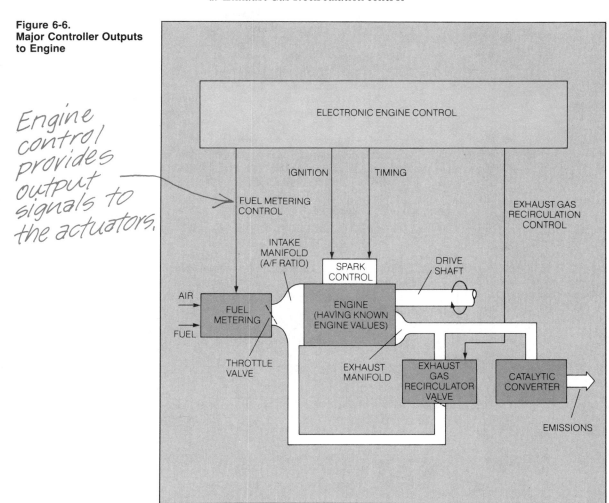

Engine control provides output signals to the actuators.

DEFINITION OF ENGINE PERFORMANCE TERMS

Common terms are used to describe an engine's performance. Here are a few:

Power

The most common performance rating that has been applied to automobiles is a power rating of the engine. It normally has been given in horsepower. Power is the rate at which the engine is doing useful work. Power varies with engine speed and throttle angle. Power may be measured at the drive wheels or at the engine output shaft depending upon which is desired. It is more convenient and useful to the designer of an electronic engine control system to know the output power of *only the engine*. This permits realistic comparisons of data as engine controls are varied. To make such measurements, an engine dynamometer is used. This dynamometer is similar to the one in *Figure 6-1* except the engine output shaft drives the dynamometer directly instead of coupling the output through wheels and rollers.

Power is a measurement of an engine's ability to perform useful work. Brake power, which is measured with an engine dynamometer, is the actual power developed by the engine minus losses due to internal friction.

The power delivered by the engine to the dynamometer is called the brake power and is designated P_b. The brake power of an engine is *always less* than the total amount of power which is actually developed in the engine. This developed power is called the indicated power of the engine and is denoted P_i. The indicated power differs from the brake power by the loss of power in the engine due to friction between cylinder and piston and other friction losses.

$$P_b = P_i - \text{friction losses}$$

BSFC

BSFC is a measurement of an engine's fuel economy. It is the ratio of fuel flow to the brake power output of the engine.

Fuel economy can be measured while the engine delivers power to the dynamometer. The engine is typically operated at a fixed RPM and a fixed brake power, P_b (fixed generator load), and the fuel flow rate, r_f (in pounds/hour), is measured. The fuel consumption is then given as the ratio of this fuel flow rate r_f to the brake power P_b output. This fuel consumption is knows as the brake specific fuel consumption or BSFC:

$$\text{BSFC} = \frac{r_f}{P_b}$$

By improving BSFC of the engine, the fuel economy of the vehicle in which it is installed will also be improved. Electronic controls help to greatly improve BSFC.

Torque

Torque is a measurement of the twisting force of an engine's crankshaft.

Engine torque is the *twisting* action which is produced on the crankshaft by the cylinder pressure pushing on the piston during the power stroke. Torque is produced whenever a force is applied to a lever. The length of the lever (the lever arm) in the engine is the "throw" of the crankshaft (the offset from the crankshaft centerline of the point where the force is applied). The torque is expressed as the product of this force and the length of the lever.

The units of torque are N.m (newton meters) in the metric system or ft. lb. (foot pounds) in the English system. One ft. lb. is the torque produced by one pound acting on a lever arm one foot long. The torque of a typical engine varies with RPM.

Volumetric Efficiency

The variation in torque with RPM is strongly influenced by the volumetric efficiency or "breathing efficiency". Volumetric efficiency actually describes how well the engine functions as an air pump, drawing air and fuel into the various cylinders. It depends upon various engine design parameters, such as, piston size, piston stroke, number of cylinders, etc.

Thermal Efficiency

Thermal efficiency expresses the mechanical energy which is delivered to the vehicle relative to the energy content of the fuel. In the typical SI engine, 35% of the energy which is available in the fuel is lost as heat to the coolant and lubricating oil, 35% is lost as heat and unburned fuel in exhaust gases, and another 10% is lost in engine and drivetrain friction. This means that only about 20% is available to drive the vehicle and accessories. Notice what a small fraction of the energy available in gasoline produces useful work.

Calibration

Setting the air/fuel ratio and ignition timing for an engine was called calibration. With new electronic control systems, calibration includes not only air/fuel ratio and ignition timing, but other variables as well.

ENGINE MAPPING

The development of any control system comes from knowledge of the "plant" or system which is to be controlled. In the case of the automobile engine, this knowledge of the plant (the engine in this case) comes primarily from a process which is called engine mapping.

For engine mapping, the engine is connected to a dynamometer and operated throughout its entire speed and load range. Measurements are made of the important engine variables while varying quantities such as the air/fuel ratio, the spark control in a known and systematic manner. Such engine mapping is done in engine test cells having complex instrumentation under computer control.

From this mapping, a mathematical model is developed which explains the influence of every measurable variable and parameter on engine performance. The control system designer must select a control configuration, control variables and control strategy which will satisfy all performance requirements (including stability) as computed from this model and which are within the other design limits such as cost, quality, and reliability. To understand a typical engine control system, let's look at the influence of control variables upon engine performance.

Effect of Air/Fuel Ratio on Performance

Figure 6-7 illustrates variation in performance variables of torque, T, and brake power, BSFC, as well as engine emissions with variations in air/fuel ratio with fixed spark timing and a constant engine speed. Recall that relatively low air/fuel ratio (below 14.7) is termed a 'rich' mixture and relatively high air/fuel ratio (above 14.7) is termed a 'lean' mixture.

Note from *Figure 6-7* that the torque, T, reaches a maximum in the air/fuel ratio range of 12 to 16. The exact air/fuel ratio for which torque is maximum depends upon the engine configuration, engine speed, and ignition timing.

**Figure 6-7.
Typical Variation of
Performance with a
Variation in Air/Fuel
Ratio**

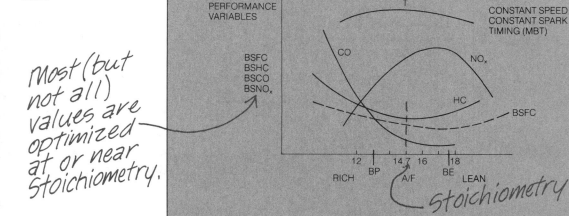

Most (but not all) values are optimized at or near Stoichiometry.

Stoichiometry

The air/fuel ratio has a significant effect on engine torque and emissions. A particular problem is that nitrous oxide emissions and engine torque are greatest near the same air/fuel ratio.

A further examination of *Figure 6-7* reveals the influence of the air/fuel ratio on the emission of the oxides of nitrogen (NO_x). In this case, however, note the relative maximum which occurs in NO_x emissions is near the air/fuel ratio where torque is greatest. As we shall see, *this characteristic of the engine causes some rather challenging problems in attempting to control exhaust emission while preserving vehicle performance.*

Effect of Spark Timing on Performance

Figure 6-8 is a typical pressure-volume diagram for a four-stroke engine. Top dead center (TDC) is at B and bottom dead center (BDC) is at A. The stroke from A to B is the compression stroke; B to C is the power stroke; C to D is the exhaust stroke and D to A is the intake stroke. The valve openings and closures are defined by points 1 to 4. Spark occurs at 5 and the flame extinguishes at 6.

**Figure 6-8.
Typical Pressure-Volume
Diagram for a Four-
Stroke Gasoline Engine**

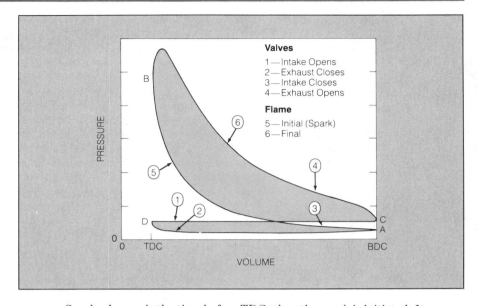

Spark timing also has a
major effect on emissions
and engine performance.
Maximum engine torque
occurs at MBT.

Spark advance is the time before TDC when the spark is initiated. It is usually expressed in number of degrees of crankshaft rotation relative to TDC. *Figure 6-9* reveals the influence of spark timing upon brake specific exhaust emissions with constant speed and constant air/fuel ratio. Note that both NO_x and HC generally increase with increased advance of spark timing. BSFC and torque are also strongly influenced by timing. *Figure 6-9* shows that maximum torque occurs at a particular advanced timing referred to as Minimum advance for Best Timing (MBT).

**Figure 6-9.
Typical Variation of
Performance with Spark
Timing**

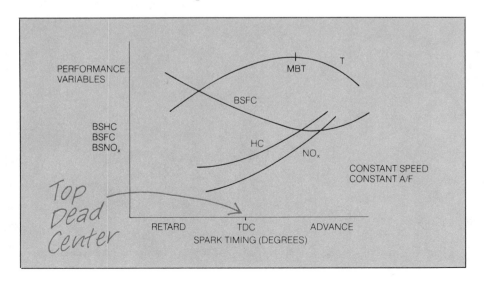

Effect of Exhaust Gas Recirculation on Performance

Exhaust gas recirculation greatly reduces nitrous oxide emissions.

Up to this point in our discussion, we have been considering the traditional calibration parameters of the engine; air/fuel and spark timing. However, by adding another calibration parameter to air/fuel and spark timing, the undesirable exhaust gas emissions of NO_x can be significantly reduced while maintaining a relatively high level of torque. This new parameter is exhaust gas recirculation (EGR), which consists of recirculating a precisely controlled amount of exhaust gas into the intake. In *Figure 6-4* we showed that exhaust gas recirculation is a major sub-system of the overall control system. Its influence upon emissions is shown in *Figure 6-10* and *6-11* as a function of the percentage of exhaust gas in the intake. *Figure 6-10* shows the dramatic reduction in NO_x emission when plotted against air/fuel ratio, while *Figure 6-11* shows the effect on performance variables as the percent of EGR is increased. Note that the emission rate of NO_x is most strongly influenced by EGR and decreases as the percentage of EGR increases. The HC emission rate increases with increasing EGR; however, for relatively low EGR percentages the HC rate changes only slightly.

The mechanism by which EGR affects NO_x production is related to the peak combustion temperature. Roughly speaking, the NO_x generation rate increases with increasing peak combustion temperature if all other variables remain fixed. Increasing EGR tends to lower this temperature; therefore, it tends to lower NO_x generation.

**Figure 6-10.
NO_x Emission as
Function of EGR at
Various A/F Ratios**

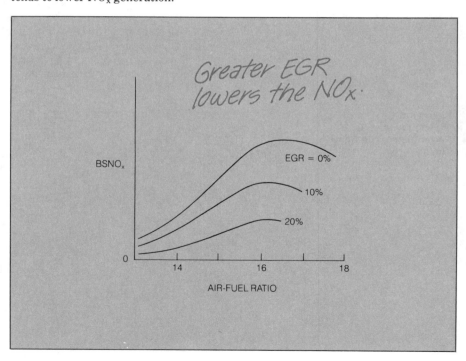

Figure 6-11.
Typical Variation of
Engine Performance with
EGR

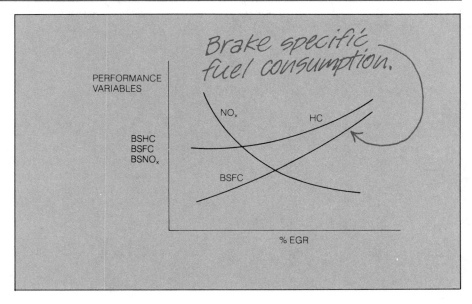

CONTROL STRATEGIES

It is the task of the electronic control system to set the calibration for each engine operating condition. There are many sets of possible control strategies to set the control variables for any given engine and each tends to have its own advantages and disadvantages. Moreover, each automobile manufacturer has a specific configuration which differs from competitive systems.

However, our discussion is about a typical electronic control system which is highly representative of all electronic control systems for engines used by U.S. manufacturers.

Catalytic Converter

This typical system is one which has a catalytic converter in the exhaust system. Exhaust gases passed through this device are chemically altered in a way which helps meet EPA standards. Essentially, the catalytic converter reduces the concentration of undesirable exhaust gases coming out of the tail pipe relative to engine-out gases (the gases coming out of the exhaust manifold).

The use of catalytic converters to reduce emissions leaving the tailpipe allows engines to be calibrated for better performance without violating emission regulations.

The EPA regulates only the exhaust gases which leave the tail pipe; therefore, if the catalytic converter reduces exhaust gas emission concentrations by 90%, then the engine exhaust gas emissions at the exhaust manifold can be about 10 times higher than the EPA requirements. This has the significant benefit of allowing engine calibration to be set for better performance than would be permitted if exhaust emissions in the engine exhaust manifold had to satisfy EPA regulations. This is the type of system that is chosen for the typical electronic engine control system.

Several types of catalytic converters are available for use on an automobile. The desirable functions of a catalytic converter include:

a. Oxidation of hydrocarbon emissions to carbon dioxide (CO_2) and water (H_2O),
b. Oxidation of CO to CO_2, and
c. Reduction of NO_x to nitrogen (N_2) and oxygen (O_2).

Oxidizing Catalytic Converter

The oxidizing catalytic converter increases the rate of oxidation of HC and CO to further reduce HC and CO emissions.

The oxidizing catalytic converter has been one of the most significant devices for controlling exhaust emissions since the era of emission control began. An oxidizing catalyst is shown in *Figure 6-12*. The purpose of the oxidizing catalyst (OC) is to increase the rate of chemical reaction, which initially takes place in the cylinder as the compressed air-fuel mixture burns, toward an exhaust gas that has a complete oxidation of HC and CO to H_2O and CO_2.

**Figure 6-12.
Oxidizing Catalytic
Converter**

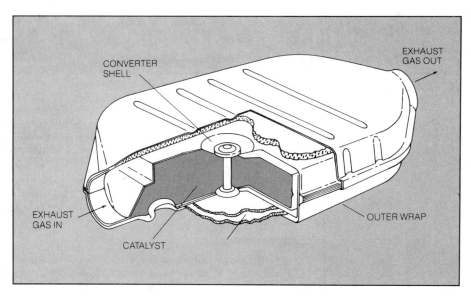

The extra oxygen which is required for this oxidation is often supplied by adding air to the exhaust stream from an engine driven air pump. This air, called secondary air, is normally introduced into the exhaust manifold immediately following the exhaust valves.

The most significant measure of the performance of the OC is its conversion efficiency n_c.

$$n_c = \frac{(M_i - M_o)}{M_i}$$

where

M_i = Mass flow rate of gas into converter.
M_o = Mass flow rate of gas leaving converter.

The conversion efficiency of the OC depends upon its temperature. *Figure 6-13* shows the conversion efficiency of a typical OC for both HC and CO as functions of temperature. Above about 300°C, the efficiency approaches 98% to 99% for CO and more than 95% for HC.

Figure 6-13.
Oxydizing Catalyst
Conversion Efficiency
Versus Temperature

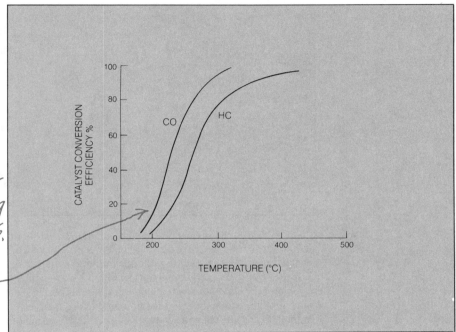

Note loss of efficiency at low operating temperatures.

The Three-Way Catalyst

The three-way catalyst uses a specific chemical design to reduce all three major emissions (HC, CO, and NO_x) by approximately 90%.

Another catalytic converter configuration which is extremely important for modern emission control systems is called the three-way catalyst (TWC). It uses a specific catalyst formulation containing platinum, palladium and rhodium which can reduce NO_x and oxidizes HC and CO all at the same time. It is called three-way because it simultaneously reduces the concentration of all three major undesirable exhaust gases by about 90%.

The conversion efficiency of the TWC for the three exhaust gases depends mostly upon air/fuel ratio. Unfortunately, the air/fuel ratio for which NO_x conversion efficiency is highest corresponds to a very low conversion efficiency for HC and CO and vice-versa. However, as shown in *Figure 6-14*, there is a very narrow range of air/fuel ratio (called the window) in which an acceptable compromise exists between NO_x and HC/CO conversion efficiencies. The conversion efficiencies within this window are sufficiently high to meet the very stringent EPA requirements established so far.

**Figure 6-14.
Conversion Efficiency
of TWC**

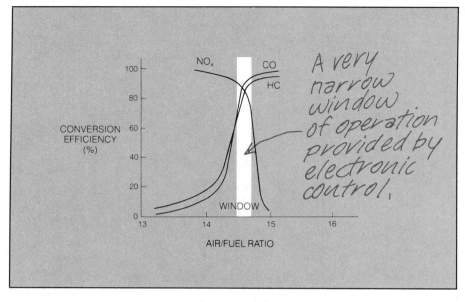

Note that this window is only about 0.1 air/fuel ratio wide (± 0.05 air/fuel ratio), and is centered at stoichiometry. Recall that stoichiometry is the air/fuel ratio that would result in complete oxidation of all carbon and hydrogen in the fuel if burning in the cylinder were perfect. For gasoline, stoichiometry corresponds to an air/fuel ratio of 14.7. This ratio and the concept of stoichiometry is extremely important in an electronic fuel controller. In fact, we will see that the primary function of most modern electronic fuel control systems is to maintain average air/fuel ratio at stoichiometry.

Control of average air/fuel ratio to the tolerances of the TWC window is beyond the capabilities of a conventional carburetor. However, as was mentioned previously, the electronic fuel control system can meet such a performance requirement. It is the primary function of the electronic engine control system.

The three-way catalyst operates at peak efficiency when the air/fuel ratio is at or very near stoichiometry. An electronic fuel control system is required to maintain the required air/fuel ratio.

ELECTRONIC FUEL CONTROL SYSTEM

For an understanding of the configuration of an electronic fuel control system, refer to *Figure 6-15*. It has been shown that an electronic fuel control system requires sensors to measure the state of the engine. It also requires one or more actuators to do the actual controlling. The sensors measure: exhaust gas oxygen, manifold absolute pressure, crankshaft angular position and speed, inlet air and coolant temperatures. Actuators are energized to control the air/fuel ratio.

Figure 6-15.
Simplified Block
Diagram of Fuel Control
System

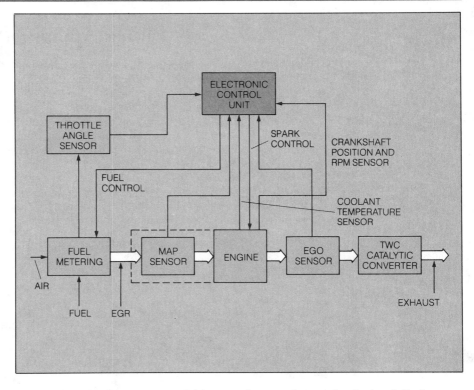

Figure 6-15.
Simplified Block
Diagram of Fuel Control
System

Depending on conditions, an electronic fuel system can operate in open loop or closed loop mode. In either mode, exhaust emissions satisfy government limits if the air/fuel ratio is held very near stoichiometry.

The primary purpose of this control system is to maintain the air/fuel ratio at or near stoichiometry. This is accomplished in two modes (during normal operation): open loop and closed loop. The concepts of these modes were discussed in Chapter 2. Recall that an error signal is fed back from output to input in the closed loop mode. The electronic fuel control system can operate closed loop only when certain conditions are satisfied. The open loop mode is employed whenever these conditions are not satisfied. However, for either mode, the exhaust emissions will satisfy EPA limits if the average air/fuel ratio is held within the tolerance limits of ± 0.05 of stoichiometry.

In addition to the open loop and closed loop control modes, a practical fuel control system has other operating modes depending on engine conditions. These handle such conditions as starting, rapid acceleration or heavy load, sudden deceleration, idling, etc. These will be covered in more detail in the next chapter.

Closed Loop Fuel Control

The step-by-step events after engine start begin with the system operating in the open loop mode. After a set of operating conditions are satisfied, the system is converted to closed loop operation. However, it is easier to understand the total system operation by beginning the discussion with an explanation of closed loop operation. The series of steps that occur in time from engine start will be discussed in Chapter 7.

In the closed loop mode of operation, the signals from the EGO sensor are used by the electronic controller to adjust the air/fuel ratio through the fuel metering actuator.

Figure 6-16 is a block diagram of a closed loop fuel control system. It operates as follows. For any given set of operating conditions, the fuel metering actuator provides fuel flow to produce an air/fuel ratio set by the controller output. This mixture is burned in the cylinder and the combustion products leave the engine through the exhaust pipe. The exhaust gas oxygen sensor (EGO) generates a signal for the controller input which depends upon the air/fuel ratio. This signal tells the controller to adjust the fuel flow rate for the required air/fuel ratio, thus, completing the loop.

**Figure 6-16.
Simplified Typical Closed
Loop Fuel Control
System**

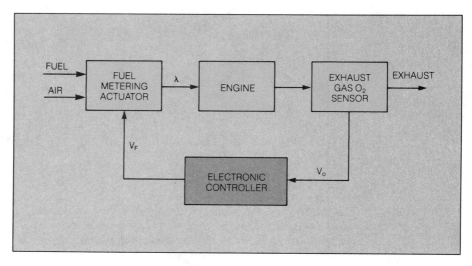

The control scheme which is most commonly employed in practice results in the air/fuel ratio cycling around the desired set point of stoichiometry. Recall that this type of control is provided by a limit cycle controller. The important parameters for this type of control include the amplitude and frequency of excursion away from the desired stoichiometric set point. Fortunately, the engine's mechanical characteristics are such that only the time average air/fuel ratio determines its performance. The variation in air/fuel ratio during the limit cycle operation is so rapid it has no effect upon engine performance or emissions provided the average air/fuel ratio remains at stoichiometry.

Fuel Metering

As shown in *Figure 6-16*, the closed loop fuel control is based upon two critical components: the fuel metering actuator and the exhaust gas oxygen sensor. First, let's discuss the actuator. In Chapter 5, the two classes of fuel metering actuators were discussed, the electronic carburetor and the throttle body fuel injector. The electronic carburetor is essentially a conventional carburetor which has been modified to permit electrical control of the metering rods.

The actuators commonly used for electronic control of fuel metering are the throttle body fuel injector and the electronic carburetor.

The TBFI typically consists of one or two solenoid operated fuel injectors which are mounted in a specially designed housing on the intake manifold. This class of TBFI functions similarly to a carburetor in that fuel is injected into and atomized by the moving air stream which flows into the intake manifold. (Recall that fuel is pulled from the conventional carburetor by the air stream moving through the venturi.) It differs from true fuel injection where fuel is injected directly into each cylinder. The calibration of the TBFI is controlled by the electronic control system.

Exhaust Gas Oxygen Concentration

The EGO sensor is used to determine the air/fuel ratio.

The second critical component, the EGO sensor, also was explained in Chapter 5. Recall that the EGO generates an output signal which depends upon the amount of oxygen in the exhaust. This oxygen level, in turn, depends upon the air/fuel ratio entering the engine. The amount of oxygen is relatively low for rich mixtures and relatively high for lean mixtures. An equivalence ratio λ (lambda) commonly used by automotive engineers is:

$$\lambda = \frac{\text{air/fuel}}{\text{air/fuel at stoichiometry}}$$

Recall that $\lambda = 1$ corresponds to stoichiometry, that $\lambda > 1$ corresponds to a lean mixture with an air/fuel ratio greater than stoichiometry and $\lambda < 1$ corresponds to a rich mixture where the air/fuel ratio is less than stoichiometry. (The EGO sensor is sometimes called a lambda sensor.)

Lambda is used in the block diagram of *Figure 6-16* to represent the equivalence ratio at the intake manifold. The exhaust gas oxygen concentration determines the EGO output, V_o. The EGO output voltage switches, as discussed in Chapter 5, abruptly between the lean and the rich levels as the air/fuel ratio crosses stoichiometry.

In a closed loop system, the time delay between sensing a deviation and the action to correct for the deviation must be compensated for in system design.

The operation of the control system using the EGO output, V_o of *Figure 6-16*, is complicated somewhat because of the time delay from the time that λ changes at the input until V_o changes at the exhaust. This time delay, t_D, is in the range of 0.1 to 0.2 second, depending on engine speed. It is the time that it takes the output of the system to respond to a change at the input. The electrical signal from the EGO sensor (V_o) going into the controller produces a controller output of V_F which energizes the fuel metering actuator.

Closed Loop Operation

The air/fuel ratio in a closed loop system is always increasing or decreasing in the vicinity of stoichiometry. This is in response to the EGO sensor's output, which indicates a rich or lean fuel mixture.

This control system operates as a limit cycle controller where the air/fuel ratio cycles up and down about the set point of stoichiometry as shown in *Figure 6-17*. The air/fuel ratio is either increasing or decreasing: it is never constant. The increase or decrease is determined by the EGO sensor output voltage. Whenever the EGO output voltage level indicates a lean mixture, the controller causes the air/fuel ratio to decrease; that is, to change in a direction of a rich mixture. On the other hand, whenever the EGO sensor output voltage indicates a rich mixture, the controller changes air/fuel ratio in the direction of a lean mixture.

The simplified waveforms of V_F, λ and V_o are shown in *Figure 6-17*. These waveforms will be used to understand the operation of the closed loop fuel control system. The time, t_D, is used as the major division of time for the time axis in *Figure 6-17*.

Let's begin by considering V_F, the output of the electronic controller. In our simplified system, this waveform, shown in *Figure 6-17a*, increases or decreases linearly with time depending upon the EGO output. Beginning with

**Figure 6-17.
Simplified Waveforms in
a Closed Loop Fuel
Control System**

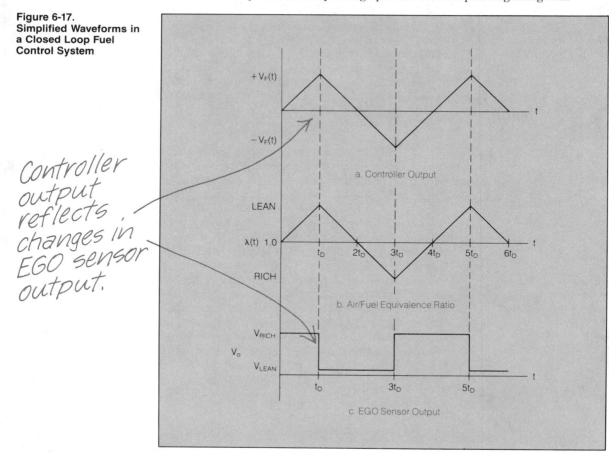

Controller output reflects changes in EGO sensor output.

$t = 0$, this voltage increases linearly with time to a maximum which occurs at t_D. Then the waveform decreases linearly reaching its lowest level at $3t_D$. The cycle repeats continuously.

The fuel metering actuator is presumed to control fuel in such a way that the intake equivalence ratio, λ, increases in proportion to V_F as shown in *Figure 6-17b.* If this were expressed as a transfer function mathematically, it would appear as:

$$\lambda - 1 = k_F V_F$$

where k_F is a constant for the fuel metering actuator. When the controller voltage equals zero ($V_F = 0$), then λ equals one ($\lambda = 1$) and the input mixture is at stoichiometry. As V_F varies, λ varies. A positive V_F produces a lean mixture (i.e., $\lambda > 1$) and a negative V_F, a rich mixture ($\lambda < 1$).

The EGO sensor output begins at time $t = 0$ and at a voltage level for a rich mixture, V_R. When the influence of the lean mixture (which takes place between $t = 0$ and $2t_D$) reaches the EGO, it switches to V_L. This occurs at a time t_D after the lean mixture first occurs. The response of the output V_o is delayed in time from the input change by the time delay, t_D.

The intake mixture is richer than stoichiometry between time $2t_D$ and $4t_D$. The effect of the rich mixture reaches the EGO at time $3t_D$ (i.e. at time t_D after the rich mixture occurs). The end of the rich mixture reaches the EGO at time $5t_D$ (t_D after the end of the input rich period at $4t_D$). This cycle repeats with a period of $4t_D$.

We should emphasize that the direction of the electronic controller output waveform *(Figure 6-17a)* changes whenever the EGO sensor output *(Figure 6-17c)* changes. The shape of the electronic controller output waveform determines the variation of λ with time. Straight line segments were used in order to simplify the explanation; however, the waveform may be curved and it may not be linear with time. The controller output voltage for the two EGO sensor output states also is shown the same amplitude, but of opposite sign. That too is for simplification because the actual controller output voltage amplitudes may not be equal. These simplifications are consistent with the simplified system block diagram and the explanation still applies to the operation under real conditions.

Frequency and Deviation of the Fuel Controller

Recall from Chapter 2 that a limit cycle controller controls a system between two limits and that it has an oscillatory behavior; that is, the control variable oscillates about the set point or the desired value for the variable. The simplified fuel controller operates in a limit cycle mode and, as shown in *Figure 6-17b*, the equivalence air/fuel ratio, λ, oscillates about stoichiometry, (i.e., average air/fuel ratio is 14.7). The two end limits are determined by the rich and lean voltage levels of the EGO sensor, by the controller, and by the characteristics of the fuel metering actuator.

When the electronic controller output voltage is zero, the air/fuel ratio is at stoichiometry. In this example, a positive voltage from the controller produces a lean mixture, while a negative voltage produces a rich mixture.

The electronic controller output waveform changes follow the EGO sensor output waveform changes. In actual practice, the changes may not be linear.

The transport delay is the time necessary for the EGO sensor to sense a change in fuel metering. As engine speed increases, the transport delay decreases.

The frequency of oscillation of this limit cycle control system, f_L, is defined as the reciprocal of its period. We can see from the waveform in *Figure 6-17b* that the period of one complete cycle is $4t_D$ (recall that t_D is the transport delay from fuel metering to EGO). Thus, the frequency of oscillation f_L is:

$$f_L = \frac{1}{4t_D}$$

expressed in hertz (cycles per second). This means that the shorter the transport delay, the higher the frequency of the limit cycle. The transport delay decreases as engine speed increases; therefore, the limit cycle frequency increases as engine speed increases. This is depicted in *Figure 6-18* for a typical engine.

**Figure 6-18.
Typical Limit Cycle
Frequency vs RPM**

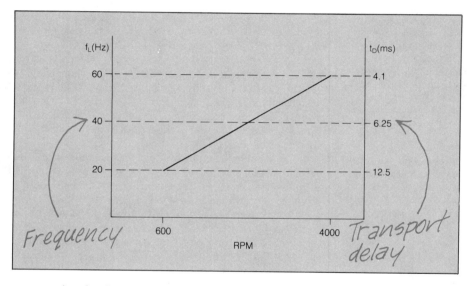

Although the air/fuel ratio is constantly swinging up and down, the average value of deviation is held to within + .05 of the 14.7:1 ratio.

Another important aspect of limit cycle operation is the deviation of λ from stoichiometry. It is important to keep this deviation small because the net TWC conversion efficiency is optimum for λ = 1. The maximum deviation of λ from unity is denoted d and given by:

$$d = (k_F)(S)(t_D)$$

where S = slope of the voltage V_F with respect to time (i.e. in volt/sec.). This deviation typically corresponds to an air/fuel ratio deviation of about ±1.

It is important to realize that the air/fuel ratio oscillates between a maximum value and a minimum value. There is, however, an average value for the air/fuel ratio which is intermediate between these extremes. Although the deviation of the air/fuel ratio during this limit cycle operation is about ±1, the *average* air/fuel ratio is held to within ± 0.05 of the desired value of 14.7.

Figure 6-19 shows a typical curve of this maximum deviation versus engine speed. Note that this deviation decreases with increasing engine speed because of the corresponding decrease in t_D. The parameters of the control system are adjusted such that d at the worst case is within the required acceptable limits for the TWC used.

Figure 6-19.
Typical Limit Cycle
Maximum Deviation from
Set Point as Speed
Varies

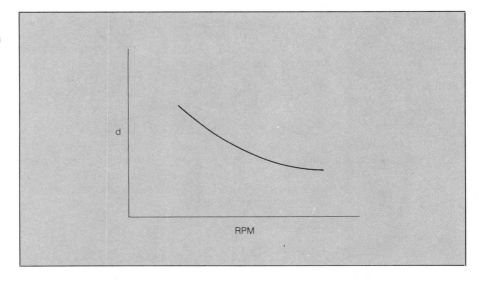

The preceding discussion applies only to a simplified idealized analog fuel control system; i.e., the system is operating with continuously changing variables between the limits and the fuel metering is changing in an analog fashion as a result of the continuously changing V_F. In the next chapter, we will explain the operation of practical electronic fuel control systems where the main signal processing is done with digital techniques. In such a system, the typical fuel metering actuator is a TBFI and it operates in a pulsed mode rather than a continuous analog mode. Also, variables other than EGO are sensed and the control computer makes several calculations to determine which actuators to control and how to control them. With this understanding of a closed loop system, let's look at the open loop mode of operation of the simplified fuel control system.

OPEN LOOP MODE

Open loop fuel control systems also must maintain the air/fuel mixture at or near stoichiometry, but these systems must do it without the benefit of feedback.

The open loop mode of fuel control must accomplish the same thing as the closed loop mode; that is, it must maintain air/fuel ratio very close to stoichiometry for efficient system operation with the TWC used. However, it must do it without feedback from the exhaust gas oxygen sensor output which determines the air/fuel ratio limits from the selected operating point.

Although the open loop mode of operation varies somewhat from one model to the next, many features of this mode of operation are common to all models. We will describe the operation of a 'typical' open loop mode control.

Measuring Air Mass

Probably the most common open loop fuel control system is based on a rather simple concept of knowing the air mass drawn into any cylinder. If this were known exactly, then the correct mass of fuel to be injected for stoichiometry would be known. That is, the ratio of air mass to fuel mass should be 14.7.

Most open loop systems operate by estimating the amount of air taken into an engine and calculating the amount of fuel needed to maintain stoichiometry. The speed-density method is commonly used to estimate the mass flow rate of air into an engine.

Unfortunately, the technology to directly measure the mass of air drawn into each cylinder has not yet been perfected. However, it is possible to closely estimate the mass flow rate of air into the engine intake by measuring other quantities. It is then theoretically possible to adjust the fuel flow rate such that the ratio of the mass flow rate of air to the mass flow rate of fuel is at stoichiometry. Most of the production fuel control systems operate essentially on this principle in the open loop mode.

Speed-Density Method

The common method of estimating mass flow rate of air into the engine is known as the speed-density method. This is based upon the concept of mass density as applied to air. As shown in *Figure 6-20a*, for a given volume of air at a temperature of T and with a mass of M_a, the density, d_a, of a sample of air occupying volume V is given by:

$$d_a = \frac{M_a}{V}$$

If M_a is in pounds and V is in cubic inches, then d_a is in pounds per cubic inch. This assumes a constant pressure to keep the air in volume V. The mass of air in a given volume depends on the temperature; therefore, the density of air depends upon its temperature. Cooler air is more dense than relatively warmer air. Tables of the density of air measured versus temperature are available and can be stored in the system as look-up tables. (Look-up tables will be discussed in Chapter 7.)

A given volume of air moving past a fixed reference point during a specific period of time is the volume flow rate.

This notion can be extended to a moving stream of air flowing through the intake of an engine as shown in *Figure 6-20b*. V, the given volume, flows by a reference point. It takes a certain period of time, t_1. Other similar volumes take the same amount of time. Therefore, the volume moving in a set time is a volume flow rate. If the volume flow rate in cubic inches per second is R_V, then the mass flow rate R_M in pounds per second is given by the product of density and volume flow rate:

$$R_M = R_V \times d_a$$

**Figure 6-20.
Volume Flow Rate**

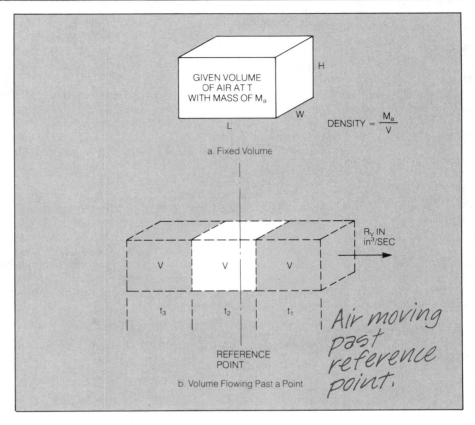

GIVEN VOLUME
OF AIR AT T
WITH MASS OF M_a

H

W

L

$DENSITY = \dfrac{M_a}{V}$

a. Fixed Volume

R_V IN
in^3/SEC

V   V   V

t_3 t_2 t_1

Air moving past reference point.

REFERENCE
POINT

b. Volume Flowing Past a Point

A close estimate of volume flow rate can be made by considering actual engine displacement and given engine speeds.

Although the technology for cost-effectively measuring volume flow rate has not yet been perfected, a relatively close estimate of R_V can be made. We know that the engine acts like an air pump during intake. If the engine were a perfect pump, it would draw in a volume of air equal to its displacement, D, for each two complete crankshaft revolutions. Then, for this ideal engine running at a speed RPM, the volume flow rate would be:

$$R_V = \left(\frac{RPM}{60}\right)\left(\frac{D}{2}\right)$$

For this ideal engine, with D known, R_V could be obtained simply by measuring RPM.

Unfortunately, the engine is not a perfect air pump. In fact, the actual volume flow rate for an engine having displacement D, running at speed RPM is given by:

$$R_V = \left(\frac{RPM}{60}\right)\left(\frac{D}{2}\right)n_V$$

where n_V = volumetric efficiency.

Volumetric Efficiency

The volumetric efficiency is a number between 0 and 1 which depends upon intake manifold pressure and RPM for all engine operating conditions. For any given engine, the value of n_V can be measured for any set of operating conditions.

A table of values of n_V as a function of RPM and intake manifold pressure, MAP, can be prepared from this data. In a digital system, the table can be stored in memory as a look-up table. By knowing the displacement of the engine, measuring the RPM and MAP, and looking up the value of n_V for that RPM and MAP, the R_V can be computed using the equation above.

Including EGR

Calculating R_V is relatively easy for a computer, but another factor must be taken into account. Exhaust gas recirculation was discussed and its purpose defined. It requires a certain portion of the charge into the cylinders be exhaust gas. Because of this, a portion of the displacement, D, is exhaust gas; therefore, the volume flow rate of EGR must be known. A valve positioning sensor on the EGR valve can be calibrated to provide the flow rate.

From this information, the volume flow rate of air, R_a, can be determined by subtracting the volume flow rate of EGR from R_V, the total cylinder charge flow rate, as follows:

$$R_a = R_V - R_{EGR}$$

Substituting the equation for R_V, the volume flow rate of air is:

$$R_a = \left[\left(\frac{RPM}{60}\right)\left(\frac{D}{2}\right)n_V\right] - R_{EGR}$$

Knowing R_a and the density, d_a, gives the mass flow rate of air, R_{am}, as follows:

$$R_{am} = R_a d_a$$

Knowing R_{am}, the stoichiometry mass flow rate for the fuel, R_{fm}, can be calculated from,

$$R_{fm} = \frac{R_{am}}{14.7}$$

It is the function of the fuel metering actuator to set the fuel mass flow rate at this desired value based upon the value of R_a. The control system continuously calculates R_{am} from R_a and d_a at the temperature involved, and generates an output electrical signal to operate the fuel metering actuator to produce a stoichiometry mass fuel flow rate. If the system is an analog system, the adjustment is continuous. If it is a typical digital control system, it completes such a measurement, computation and control signal generation about 100 to 300 times each second.

ELECTRONIC IGNITION

The engine ignition system exists solely to provide an electric spark which ignites the mixture in the cylinder. In the first chapter of this book, the basics of a conventional purely electrical ignition system were discussed. The ignition system in most modern automobiles is electronic as opposed to this conventional system.

Electronic ignition has a relatively long history compared, for example, with the fuel control system. It was one of the first non-entertainment electronic systems on the automobile. Electronic ignition can either be a separate system, independent of the engine control system, or it can be incorporated as a secondary function of the engine control system. Let's discuss the separate system first.

Electronic ignition can operate as an independent system, or as a function of an integrated engine control system.

Separate System

As a separate system, the electronic ignition system is essentially an improvement of the conventional system. One of the major differences is the replacement of the breaker points with an electronic circuit (Chapter 5). An ignition timing sensor "measures" the engine angular position, i.e. the position at which the spark is to occur.

Generating the Pulse

The sensor generates a pulse which triggers the ignition electronic circuit which in turn drives the coil primary. This circuit, when so triggered, switches off the current in the coil primary thereby initiating the spark.

The concept of an engine position sensor used as an ignition timing sensor was introduced in Chapter 5. Another more detailed example is shown in *Figure 6-21*.

One kind of separate electronic ignition system uses a magnetic position sensor to determine timing of the spark. The pickup coil produces a high control voltage each time one of the cogs on the rotating ferromagnetic element passes by.

Here a permanent magnet couples to a ferromagnetic element which is mounted on the distributor shaft and rotates with it. As this element rotates, the strength of the magnetic field varies, being largest when the air gap is smallest. The time varying magnetic field induces a voltage in the coil which is proportional to the rate of change of the magnetic field and which has a waveform as illustrated in *Figure 6-22*. Each time one of the cogs on the ferromagnetic wheel passes under the coil axis, one of the sawtooth shaped pulses is generated. This wheel has one cog for each cylinder and the voltage pulses correspond to the spark time for the corresponding cylinder. In a sense, the rotary wheel plays a role similar to the cam in a conventional distributor.

**Figure 6-21.
Magnetic Position
Sensor**

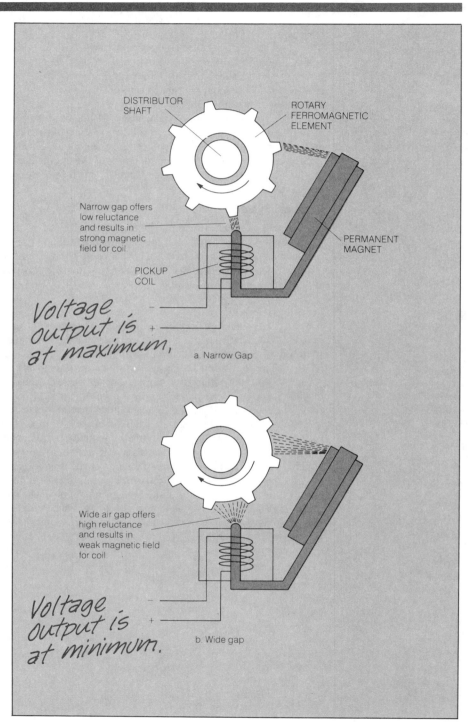

DISTRIBUTOR
SHAFT

ROTARY
FERROMAGNETIC
ELEMENT

Narrow gap offers
low reluctance
and results in
strong magnetic
field for coil.

PERMANENT
MAGNET

PICKUP
COIL

*Voltage
output is
at maximum.*

a. Narrow Gap

Wide air gap offers
high reluctance
and results in
weak magnetic field
for coil.

*Voltage
output is
at minimum.*

b. Wide gap

**Figure 6-22.
Idealized Waveform at
the Output of the Pickup
Coil**

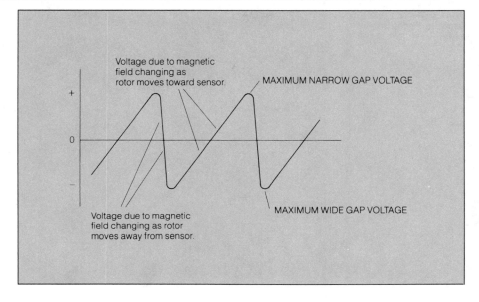

The Actuator

The pickup coil output is amplified by a control circuit, and the amplified and shaped signal controls an electronic switch. The switch controls current flow in the ignition coil circuit to generate a high voltage for the spark.

The remainder of the electronic ignition system is illustrated in *Figure 6-23* in block diagram form. The sensor output operates an ignition circuit (an actual circuit was shown in *Figure 5-40*) which generates a pulse at the correct time for ignition. This pulse is amplified and triggers a switching amplifier causing it to stop conducting. This is analogous to opening the breaker points in a conventional ignition system. The remaining action of this system is identical to the conventional system. When the primary of the coil is suddenly open circuited, the collapsing magnetic field generates a high voltage in the coil secondary. As explained in Chapter 1, this secondary voltage is connected to the rotor of the distributor. The distributor shaft is connected by gears to the camshaft and rotates synchronously with the crankshaft. The distributor connects the coil high voltage alternately to the appropriate spark plug wire.

In many cases, the coil secondary voltage is 30,000 to 35,000 volts in order to ignite the air/fuel mixture under the engine operating conditions that must prevail to control the emissions and fuel economy and still allow for spark plug and distributor variations. Extreme caution should be taken when working around this high secondary voltage. Also, a different type of insulation is used on the spark plug wires. Typically, this insulation is more likely to be damaged by improper handling than that used on conventional systems. Care also should be taken to ensure the spark plug wires are routed and clamped as originally installed by the manufacturer to protect the wires and to prevent crossfiring.

**Figure 6-23.
Electronic Ignition
System**

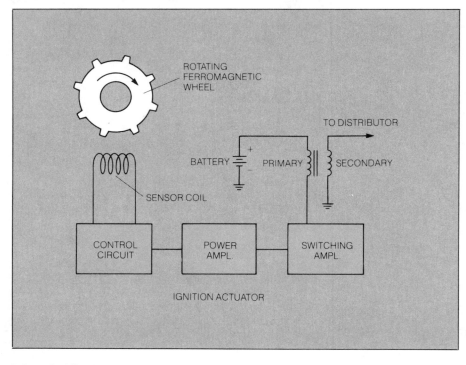

In integrated engine control systems, spark advance is controlled as a function of inputs from MAP and engine speed sensors.

Integrated System

In this and previous chapters, it was shown that the timing of the spark relative to the time the piston reaches TDC is critical, and that the spark advance markedly affects engine exhaust emissions and performance. In many of the earliest electronic ignition systems, spark advance control for changes in engine load and speed was achieved in much the same way as in a conventional system. That is, manifold vacuum and centrifugal force were employed to move the sensor coil relative to the ferromagnetic wheel by rotating a plate in the distributor upon which the sensor coil was mounted.

With a digital system controller, the spark advance can be controlled electronically. Sensors that measure MAP and engine RPM provide the inputs to the computer which calculates the spark advance needed. Then the computer generates the pulse to drive the ignition actuator circuit or to open the coil primary circuit itself. Such a system is discussed in the next chapter.

SUMMARY

In this chapter, electronic engine control systems have been discussed in order to understand the basic concepts and fundamentals. With this knowledge we will proceed to the next chapter, getting more specific by describing practical electronic engine control systems that use digital techniques.

Quiz for Chapter 6

1. What is the primary motivation for engine controls?
   a. consumer demand for precise controls
   b. automotive industry's desire to innovate
   c. government regulations concerning emissions and fuel economy

2. What is the primary purpose of fuel control?
   a. to maximize fuel economy
   b. to minimize exhaust emissions
   c. to optimize catalytic converter efficiency
   d. to maximize engine torque

3. What is the primary purpose of spark timing controls?
   a. to maximize fuel economy
   b. to minimize exhaust emissions
   c. to optimize catalytic converter efficiency
   d. to optimize some aspect of engine performance (e.g. torque)

4. What does exhaust gas recirculation do?
   a. improve fuel economy
   b. reduce NO_x emission
   c. increase engine torque
   d. provide air for the catalytic converter

5. What does secondary air do?
   a. dilute the air fuel ratio
   b. help oxidize HC and CO in the exhaust manifold
   c. help oxidize NO_x and CO in the catalytic converter
   d. help reduce the production of NO_x

6. What is air fuel ratio?
   a. the mass of air in a cylinder divided by the mass of fuel
   b. the volume of air in a cylinder divided by the volume of fuel
   c. the ratio of the mass of HC to mass of NO_x

7. What electronic device is used in engine controls?
   a. AM radio
   b. catalytic converter
   c. microcomputer

8. What air/fuel ratio is desired for a 3-way catalytic converter?
   a. 12:1
   b. 17:1
   c. 14.7:1
   d. none of the above

9. What is the desired operation of a catalytic converter on HC emissions?
   a. oxidation to H_2O and CO_2
   b. reduction to H and C
   c. reaction with NO_x
   d. none of the above

10. What is the desired operation of a catalytic converter on NO_x emissions?
    a. reaction with HC
    b. oxidation to N_2 and O_2
    c. reduction to N_2 and O_2
    d. none of the above

11. What is stoichiometry?
    a. a very lean air/fuel ratio
    b. a very rich air/fuel ratio
    c. an air/fuel ratio for which complete combustion is theoretically possible
    d. none of the above

12. How is CO emission affected by air/fuel ratio?
    a. it generally decreases with increasing air/fuel ratio
    b. it increases monotonically with air/fuel ratio
    c. it is unaffected by air/fuel ratio
    d. none of the above

13. What is MBT?
    a. mean before top-center
    b. miles per brake torque
    c. it is a spark advance angle for maximum torque
    d. none of the above

14. What is the function of electronic fuel
control in a vehicle having a
3-way catalyst?
  a. to maximize brake specific fuel
  consumption
  b. to maintain the average air/fuel
  ratio at stoichiometry
  c. to always keep the air/fuel ratio
  within ± 0.05 of stoichiometry
  d. to minimize NO_x emissions

15. What is the fuel flow rate for an
electronic fuel control system for a
vehicle having a 3-way catalyst?
  a. $R_{fm} = R_{am}/14.7$
  b. $R_a/14.7$
  c. $R_v – R_{EGR}/14.7$
  d. none of the above

16. What is one difference between a
conventional and an electronic ignition
system?
  a. there are no differences
  b. the electronic system produces a
  lower coil secondary voltage
  c. the coil is eliminated in the
  electronic system
  d. distributor points are replaced by a
  crankshaft position sensor and
  electronic circuit

17. What engine quantities are measured
to determine spark advance for an
electronic ignition system?
  a. manifold pressure and RPM
  b. coolant temperature and mass
  air flow
  c. manifold position and crankshaft
  position
  d. none of the above

18. In an electronic fuel control system,
what causes the time delay between
fuel metering and the EGO sensor
response?
  a. dynamic response of the electronics
  b. transport time of the air and fuel
  through the engine
  c. limit cycle theory
  d. none of the above

19. Brake power of an engine is:
  a. the power required to decelerate
  the car
  b. an electronic system for stopping
  the car
  c. the difference between indicated
  power and power losses in
  the engine
  d. none of the above

20. What is engine calibration?
  a. adjustment of air/fuel ratio, spark
  timing, and EGR
  b. instrumentation parameter setting
  c. electronic control system
  parameters
  d. none of the above

Typical Digital Engine Control System

ABOUT THIS CHAPTER

In the last chapter, we discussed engine control systems in a general way. In this chapter, we will apply this generalized information to see how a specific digital electronic engine control system works. Before we get into the details, let's quickly review how sensor outputs are converted to digital input signals that are acceptable by the controller, how the controller can use the digital signals, and how the controller generates output signals to control actuators.

CONTROLLER SIGNALS

Controller Inputs

Control variables are represented by electrical signals from sensors. The electrical signal, as discussed in Chapter 5, is either a voltage or current. It might be an analog voltage continuously varying with time, a pulse, or a constant binary level voltage. These will be examined briefly to understand how an electronic controller uses these signal inputs. *Figure 7-1* shows the type of electrical signals that the electronic controller normally accepts.

Figure 7-1a is a binary logic level signal. The electrical signal is normally at the high level (HL) or at the low level (LL). In positive logic, the high level represents a "1" and the low level a "0" in binary code. The controller looks at the signal and recognizes it as a one or a zero. The "1" could mean a switch contact on the engine is closed; a "0" could mean the switch is open.

Figure 7-1b shows four input lines from separate control variables, each with a binary logic level code. The controller may combine the signals from these four variables and form a code from them. For example, if the four inputs are sampled at different times t_1, t_2, t_3 and t_4, the controller can store the signals and form a 4-bit binary code. The signal from V_0 would be bit 0; the signal from V_1 would be bit 1; the signal from V_2 would be bit 2, and the signal from V_3 would be bit 3. All four signals when examined in time sequence by the controller form the code 1010. This code could be a particular memory address or be recognized by a logic network inside the controller to trigger a particular action. Of course, more than four variables could be sampled by using more than four bits in the code. The controller just needs a little extra time to scan all inputs.

Binary logic signals are either at a high level, representing a 1, or at a low level, representing a 0. Electronic controllers can examine individual binary signals, or a group of binary signals representing a code.

Figure 7-1.
Controller Input Signals

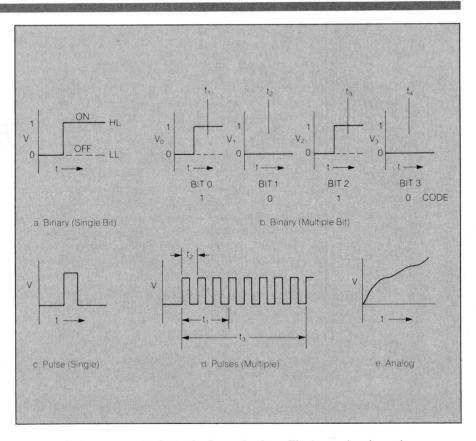

Many different forms of digital or analog signals.

a. Binary (Single Bit)

b. Binary (Multiple Bit)

BIT 0   BIT 1   BIT 2   BIT 3
1   0   1   0   CODE

c. Pulse (Single)

d. Pulses (Multiple)

e. Analog

Binary signals in the form of pulses can be used to trigger controller functions, or they can be counted by the controller to measure various timing functions.

An input may also be in the form of pulses. The input signal may be a single pulse, as shown in *Figure 7-1c*, occurring at a particular time to trigger a circuit inside the controller, or it may be a series of pulses occurring at a particular frequency as shown in *Figure 7-1d*. The controller may count the pulses in order to see how many occur in a particular time, t_1. This could be used to determine RPM of the engine, as an example. Or the controller might examine the time t_2 between two pulses to determine a time delay. Or the controller may count a given number of pulses to measure a time interval such as t_3.

A very common type of voltage from a sensor is shown in *Figure 7-1e*. This is an analog voltage. It is a continuous output voltage that varies in level with the physical quantity it represents. Since digital electronic circuitry is used in the controller, this analog voltage must be converted to a digital code so the controller can recognize it.

Analog to Digital Conversion

Analog to digital converters change an analog voltage into a binary code. Each specific analog voltage level can be represented by a different binary code.

Chapter 4 contained a general discussion of analog to digital converters. *Figure 7-2* describes how this is accomplished in more detail. The analog voltage, V_1, represents a temperature. The analog voltage V_1 is sampled by the analog to digital converter and converted into a binary code. In this case, it is an 8-bit code so that 255°F is the full-scale reading. The binary code for the voltage V_1 converts into the decimal value of 130°F. (The method for converting binary to decimal was shown in Chapter 3.) The controller uses the binary code to interpret that the sensor has detected a temperature of 130°F.

**Figure 7-2.
Analog-to-Digital
Converter**

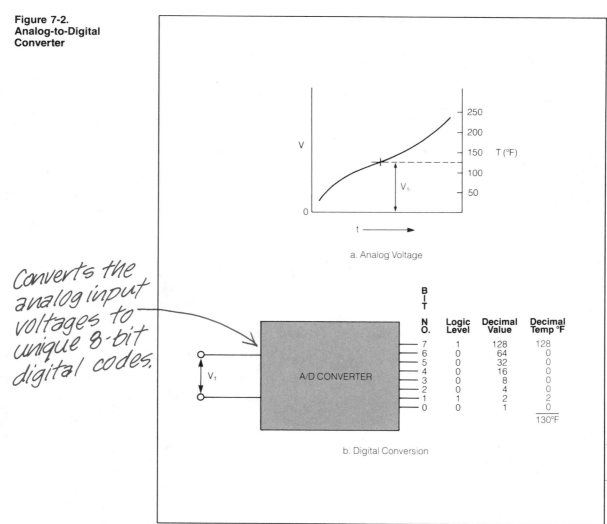

a. Analog Voltage

Converts the analog input voltages to unique 8-bit digital codes.

B I T N O.	Logic Level	Decimal Value	Decimal Temp °F
7	1	128	128
6	0	64	0
5	0	32	0
4	0	16	0
3	0	8	0
2	0	4	0
1	1	2	2
0	0	1	0
			130°F

A/D CONVERTER

V_1

b. Digital Conversion

Look-Up Table

The controller can compare binary codes produced by the ADC to values in a look-up table stored in the computer by using an exclusive OR logic gate.

Figure 7-3 shows how the controller could use this code in a look-up table that is stored in the computer. Suppose the controller has a look-up table to determine warm-up time of the engine for various engine temperatures. The sensor says the engine temperature is 130°F. The controller stores the code for 130°F in a temporary storage register R2. The controller then compares the stored code in R2 with each of the stored table values by comparing each bit of the codes, bit by bit, using an exclusive OR logic gate and storing the comparisons in R1, bit by bit. The exclusive OR logic gate has the property to have an output of 1 if either of the inputs is 1. If both inputs are 1 or both inputs are 0, then the output will be 0. Therefore, when the codes are the same, each comparison, bit by bit, will yield an output code from the exclusive OR gate that has 0 in each bit position in register R1. The controller then knows that the correct temperature code has been located in the look-up table when register R1 has a 0 in all bit positions. When this occurs, the controller uses the warm-up time code in the table corresponding to the matched temperature code. For the 130°F in this example, the warm-up code is 0110. This code might represent a warm-up time of six minutes if the binary code is used in the control system directly. Only the look-up table codes for each 10° of temperature are shown in *Figure 7-3*. Actually, there are codes for each 1° and the A/D converter only outputs codes for every 1° change in temperature.

When the compared codes match, the controller uses a corresponding control code to execute a control function. In the example, an engine temperature code translates to a control code for six minutes of engine warmup time.

**Figure 7-3.
Comparison to Locate
Value in Look-Up Table**

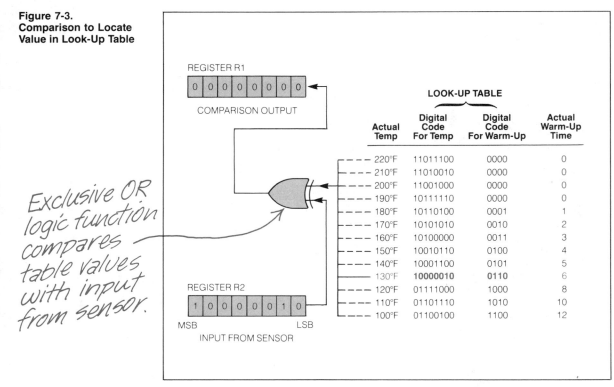

Actual Temp	Digital Code For Temp	Digital Code For Warm-Up	Actual Warm-Up Time
220°F	11011100	0000	0
210°F	11010010	0000	0
200°F	11001000	0000	0
190°F	10111110	0000	0
180°F	10110100	0001	1
170°F	10101010	0010	2
160°F	10100000	0011	3
150°F	10010110	0100	4
140°F	10001100	0101	5
130°F	10000010	0110	6
120°F	01111000	1000	8
110°F	01101110	1010	10
100°F	01100100	1100	12

Exclusive OR logic function compares table values with input from sensor.

Controller Outputs

Typical signals that the controller sends out to control actuators or other electronic circuits are shown in *Figure 7-4*. The amplitudes of the signals are shown in common binary logic levels of 1 and 0 for each case, but the signals could represent a voltage or current of any amplitude.

**Figure 7-4.
Controller Output
Signals**

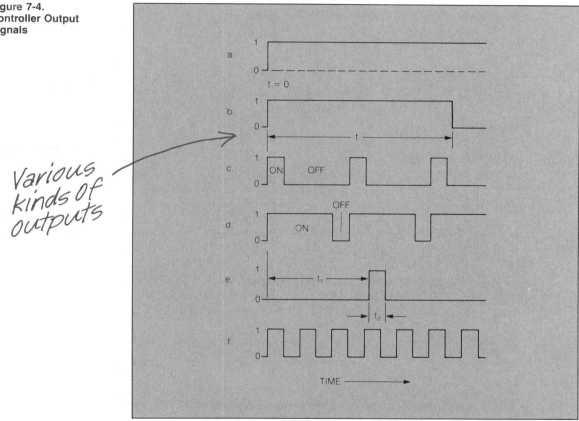

*Various
kinds of
outputs*

Controller outputs, in the form of continuous or pulsed binary signals, can trigger actuators for continuous or varying operation.

The solid line in *Figure 7-4a* is a signal to turn on an actuator at time $t = 0$ and hold it on continuously. The dotted line would be the signal for turning off at $t = 0$. *Figure 7-4b* is a timed pulse which energizes an actuator for time t. *Figure 7-4c* and *7-4d* are varying duty cycle pulses like the ones used for the TBFI fuel metering; c has a short on time and long off time, d is just the reverse.

Figure 7-4e is a single pulse of pulse width t_2 which occurs at time t_1 after $t = 0$. In some cases, only one of the times may be important. The single pulse might be used for triggering an actuator; e.g., it might be the ignition pulse to drive the electronic ignition circuit actuator or a timing pulse to synchronize other outputs from other electronic circuits.

Another example of a signal used for timing is shown in *Figure 7-4f*. It is a continuous train of pulses at a constant frequency. It is called a clock signal. Timed events that need to be synchronized are either timed directly from this output or from additional timing signals derived from it.

SYSTEM SPECIFICATIONS

Now let's see what happens in a typical digital electronic control system. Here are the specifications for the system:

1. The system will use digital techniques
2. A TWC catalytic converter will be used.
3. The average air/fuel ratio will be held to the stoichiometric ratio of 14.7 ± 0.05 for optimum conversion of HC, CO and NO_x.
4. EGR is used to cool peak combustion temperatures to reduce formation of NO_x.
5. Spark timing is controlled by a schedule (e.g., to maximize engine torque).
6. Secondary air will be routed to the catalytic converter to improve its efficiency.
7. Major control will be by closed loop where intake air/fuel ratio is detected by an EGO sensor in the exhaust manifold.
8. Open loop control of air/fuel ratio is by the speed-density method.
9. Functions to be controlled are fuel, spark timing, EGR, secondary air management and control mode selection.

The major subsystems of the engine control system are fuel control, spark control and exhaust gas recirculation control. However, because of the flexibility of digital control systems using microprocessors and microcomputers, other functions can be added. Each of the major subsystems will be discussed separately and then the full integrated system will be linked together. The first subsystem is the fuel-control system.

CONTROL MODES FOR FUEL CONTROL

An automotive engine has various operating modes as the operating conditions change. Programmed into the microcomputer in the engine control system or the fuel-control system, in this case, is control mode selection logic which determines the operating mode from the engine conditions that exist. From these engine conditions, the control system functions to be performed within each operating mode also are determined.

Engines have different modes of operation as the operating conditions change. Six different modes of operation commonly affect fuel control.

There are six different engine operating modes which affect fuel control; engine crank, engine warm up, open loop control, closed loop control, hard acceleration, deceleration and idle. The program for mode control logic determines the engine operating mode by reading various sensors and timers.

During engine crank and engine warmup modes, the controller holds the air/fuel ratio to a purposely low value (rich fuel mixture).

When the ignition key is switched on initially, the mode control logic automatically selects an engine start control scheme which provides the low air/fuel ratio required for starting the engine. Once the engine RPM rises above the cranking value, the controller identifies the engine started mode and passes control to the program for the engine warm up mode. This operating mode keeps the air/fuel ratio low to prevent engine stall during cool weather until engine coolant temperature rises above some minimum value.

After warmup, the controller switches to open loop control until accurate readings can be obtained from the EGO sensor. The controller then changes to, and remains in, closed loop mode under ordinary driving conditions.

When the coolant temperature rises, the mode control logic directs the system to operate in the open loop control mode until a certain time has elapsed and the EGO sensor warms up enough to provide accurate readings. This condition is detected by monitoring the EGO sensor's output for voltage readings above a certain minimum air/fuel rich mixture voltage set point. When the sensor has indicated rich at least once, and after the engine has been in open loop for a specific time, the control mode selection logic selects the closed loop mode for the system. The engine remains in the closed loop mode until either the EGO sensor cools and fails to read a rich mixture for a certain length of time, or a hard acceleration or deceleration occurs. If the sensor cools, the control mode logic selects the open loop mode again.

During conditions of hard acceleration or deceleration, the controller adjusts the air/fuel ratio as needed for those conditions. During idle periods, the controller adjusts engine speed to reduce engine roughness and stalling.

During hard acceleration or heavy engine load, the control mode selection logic chooses a scheme which provides a rich air/fuel mixture for the duration of the acceleration or heavy load. This scheme provides maximum torque, but poor emissions control and poor fuel economy regulation. After the need for enrichment has passed, control is returned to either open loop or closed loop depending on the control mode logic selection conditions that exist at that time.

During periods of deceleration, the air/fuel ratio is increased to reduce emissions of HC and CO due to unburned excess fuel. When idle conditions are present, control mode logic passes system control to the idle speed control mode. In this mode, the engine speed is controlled to reduce engine roughness and stalling which might occur because the idle load has changed due to air conditioner compressor operation, alternator operation, or gearshift positioning from park/neutral to drive.

Engine Crank

While the engine is being cranked, the fuel control system must provide an intake air/fuel ratio of anywhere from 2/1 to 12/1, depending on engine temperature. Low temperatures affect the carburetor's ability to atomize or mix the incoming air and fuel. At low temperatures, the fuel tends to form into large droplets in the air which don't burn as efficiently as tiny droplets. The larger fuel droplets tend to increase the apparent air/fuel ratio because the amount of useable fuel in the air is reduced; therefore, the carburetor must provide a decreased air/fuel ratio to provide the engine with a

During engine crank, the controller compares the value from the coolant temperature sensor with values stored in a look-up table to determine the correct air/fuel ratio at that temperature.

more combustible air/fuel mixture. A diagram of the system operation is shown in *Figure 7-5*. The engine temperature is read by the computer through an analog to digital converter from a temperature sensor in the engine water coolant. A look-up table is used to determine the proper air/fuel ratio at that temperature. The air/fuel ratio is determined and controlled as in the open loop mode which will be discussed in a minute. The main control concern is for reliable engine start, not for emission control or for fuel economy.

The fuel metering actuator in this case is a throttle body fuel injector (TBFI) like the one discussed in Chapter 5. The output of the controller is a variable duty cycle pulse to meter the correct amount of fuel to obtain the calculated air/fuel ratio.

**Figure 7-5.
Engine Crank Operating
Mode**

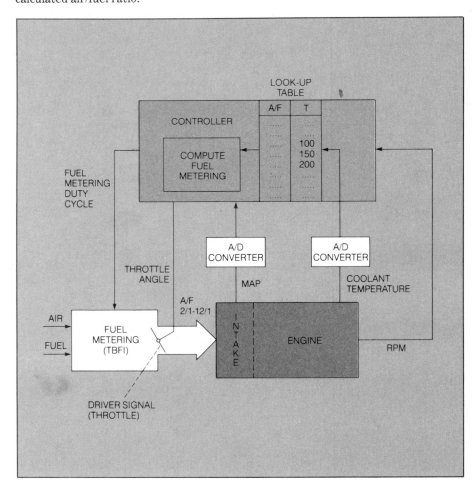

Engine Warm Up

The controller selects a warm up time from a look-up table based on the temperature of the coolant. During engine warmup, the air/fuel ratio is still rich, but it is changed by the controller as the coolant temperature increases.

While the engine is warming up, an enriched air/fuel ratio is still needed to keep it running smoothly, but the required air/fuel ratio changes as the temperature increases. Therefore, the fuel control system stays in the open loop mode, but the air/fuel ratio commands continue to be altered due to the temperature changes. The emphasis in this control mode is on rapid and smooth engine warm up. Fuel economy and emission control are still a secondary concern. A diagram of just the controller is shown in *Figure 7-6*. The controller determines the warm-up time period based on the coolant temperature when the warm-up mode was selected. Of course, an initially cold engine requires a longer warm-up time than a warm engine. The time allowed by the controller timer is chosen from a look-up table and is as short as possible so that the controller can begin full regulation of emissions and fuel economy as quickly as possible.

Figure 7-6.
Warm-Up Operating Mode

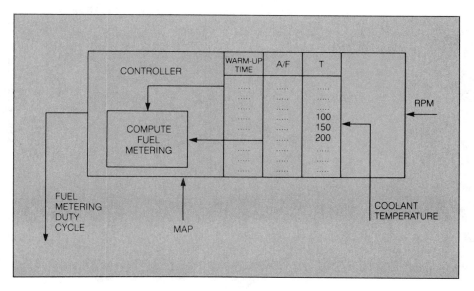

Open Loop Control

After engine warmup, open loop control is used. The most popular method uses the mass density equation to calculate the amount of air entering the intake manifold.

Once the engine is warmed up and the warm up timer has timed out, the fuel control system operates as an open loop control system to more closely control emissions and fuel economy. As mentioned in Chapter 6, the open loop fuel mass flow rate is computed from the air mass flow rate. The fuel mass flow rate is regulated so as to maintain the air/fuel ratio as close as possible to 14.7

The most popular method of computation of air mass flow rate is the mass density equation. As discussed previously, this method requires a knowledge of an engine's volumetric efficiency, cylinder displacement, and the intake air density. These parameters along with engine RPM and manifold absolute pressure (MAP) are sufficient to compute the mass flow rate of air drawn through the intake manifold.

Volumetric efficiency and cylinder displacement are well known by the engine designers for a particular engine and can be tabulated in a look-up table using engine RPM as a look-up index. Intake air density can be computed from an engine off measurement of manifold pressure (for barometric pressure) and temperature (coolant temperature). (Some systems have separate sensors to measure barometric pressure and intake air temperature directly.) A diagram of the control system operation in shown in *Figure 7-7.*

**Figure 7-7.
Open Loop Control**

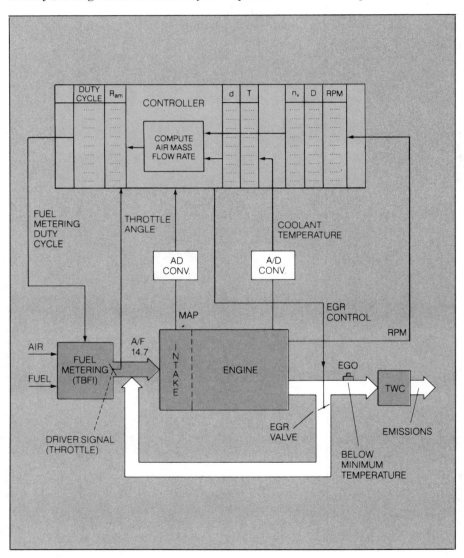

Basic air-volume flow rates, corresponding to engine RPM, are stored in a look-up table. After adjustments for MAP, air temperature, and EGR, the value is used by the controller to determine proper fuel metering.

The air volume flow-rate is computed from the look up table determined by RPM. This value is adjusted to account for MAP (barometric pressure) and air density at the air temperature that exists. The resulting value is an estimate of air mass flow rate which is then corrected for EGR mass and is used in another look up table to determine the appropriate duty cycle for fuel metering to provide the proper air/fuel ratio.

Basically there are two ways to compute the TBFI duty cycle from the mass density equations that were given in Chapter 6. The most direct way is to program the computer to perform all of the multiplications and divisions indicated in the air mass equations and then compute EGR mass using a similar equation. The EGR mass is then subtracted from the air mass and the result is used to compute the fuel mass and corresponding TBFI duty cycle. This might require 5 or 6 multiplications, 2 or 3 divisions and a few additions or subtractions. Multiplication and division in a microcomputer require 15 to 30 microseconds and addition and subtraction require 2 or 3 microseconds. After each operation is performed, the computer must check for overflow or underflow and make corrections when necessary. These computations can stretch the total air/fuel ratio computation time to several milliseconds. The computation time is very critical because a new air/fuel ratio computation must be made every 20 to 50 milliseconds and the computer has other jobs to do besides just computing air/fuel ratio.

To reduce the amount of computation time necessary for complete mass density equations, look-up tables are often used. However, direct computation of the equations also are used in some engine control systems.

To reduce computation time, look-up tables like those shown in the example of *Figure 7-7* are sometimes used to eliminate some of the multiplications and divisions. Look-up tables also reduce the overflow detection and correction problems. Input values that would normally cause overflow to occur in the direct computation approach automatically return overflow corrected values from the look-up table. Both methods, direct computation and look-up tables, are used to some extent in virtually every current production engine control system.

As before, MAP and coolant temperature are read into the computer through A/D converters. The MAP and coolant temperature are analog signals while the RPM is determined from ignition timing pulses. EGR is usually controlled by the computer so the amount of EGR is already known. In systems where EGR is not under computer control, the amount of EGR can be determined by measuring the amount of opening in the EGR valve with a sensor similar to a throttle position sensor (proportional EGR) or measuring the duty-cycle in a pulsed EGR system.

Closed Loop Control

When the coolant temperature sensor indicates that the engine is warm and the EGO sensor is providing accurate signals, the system changes to closed loop control. The controller acts as a limit cycle controller, varying fuel metering in response to the lean or rich indication from the EGO sensor.

Closed loop fuel control is selected when the engine is warm and the exhaust gas oxygen sensor has exceeded its minimum operating temperature. The intake air/fuel ratio is controlled in a closed loop by measuring the EGO at the exhaust manifold and altering the input fuel flow rate with a TBFI fuel metering actuator to correct for a rich or lean mixture indication. The EGO signal is a digital signal with two states as discussed in Chapter 6 (*Figure 6-17*). The high level indicates a rich mixture for air/fuel ratio and a low level a lean mixture. The signal is amplified by an operational amplifier and fed into a digital input port on the computer. The computer determines which side of stoichiometry the air/fuel ratio is on based on the state of the EGO signal. The full fuel control system operates as a limit cycle closed loop control system as shown in *Figure 7-8*. In this case, the fuel metering actuator is a throttle body fuel injector.

Figure 7-8.
Closed Loop Control

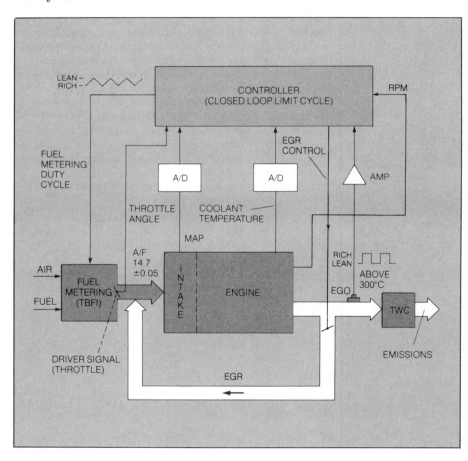

At the time that the EGO sensor detects a rich or lean condition, the TBFI duty cycle signal is reduced or increased by a fixed amount based on engine temperature, load and throttle angle. This TBFI duty cycle signal is then changed to follow the V_F signal that is ramped downward or upward at a fixed rate as discussed in Chapter 6 (*Figure 6-17*) to maintain an average air/fuel ratio within the window around the set point of stoichiometry.

Variations in engine transport delay with RPM are corrected by reducing the cycle frequency and ramp rate with decreasing RPM. The long intake to exhaust transport delay time at very low RPM tends to cause the air/fuel ratio to swing wildly between very rich and very lean conditions. Slowing the duty cycle ramp rate tends to reduce the amplitude of the swing by allowing the EGO sensor more time to react to input air/fuel ratio changes at low RPM. This keeps the average intake air/fuel ratio within acceptable limits.

Acceleration Enrichment

During hard acceleration or heavy engine load, the controller responds to higher MAP sensor values or throttle angles by increasing the duty cycle of the fuel metering. The richer air/fuel mixture provides increased engine torque.

During periods of heavy engine load such as during hard acceleration, fuel control is adjusted to provide an enriched air/fuel ratio to maximize engine torque while neglecting fuel economy and emissions.

The computer detects this condition by reading the throttle angle sensor voltage or from the MAP sensor. High intake manifold pressure or throttle angle corresponds to heavy engine load. The fuel control system controller responds by increasing the duty cycle of the fuel metering signal for the duration of the heavy load. This enrichment enables the engine to operate with a torque greater than that allowed when emissions and fuel economy are controlled within specifications.

Deceleration Enleanment and Idle Speed Control

Sudden decreases in MAP sensor values or throttle angles, indicating deceleration, result in the controller reducing the fuel metering duty cycle. The resultant lean fuel mixture reduces emissions and provides increased fuel economy.

During periods of light engine load and high RPM, such as during coast or hard deceleration, the engine requires a very lean air/fuel ratio to reduce excess emissions of HC and CO. Deceleration is indicated by a sudden decrease in MAP and throttle angle. When these conditions are detected by the control computer, it computes a decrease in the duty cycle of the fuel metering signal. The fuel may even be turned off completely for very heavy deceleration.

Idle speed control is used by some manufacturers to prevent engine stall during idle. The goal is to allow the engine to idle at as low an RPM as possible, yet keep the engine from running rough and stalling when power take-off accessories such as air conditioning compressors and alternators turn on.

When the throttle angle reaches its closed position and engine RPM falls below a preset value, the controller switches to idle speed control. A stepping motor opens a valve, allowing a limited amount of air to bypass the closed throttle plate.

The control mode selection logic switches to idle speed control when the throttle angle reaches its zero (completely closed) position and engine RPM falls below a minimum value. Idle speed is controlled by using an electronically controlled throttle bypass valve as shown in *Figure 7-9* which allows air to flow around the throttle plate and produces the same effect as if the throttle had been slightly opened. The valve is controlled in a stepper fashion; that is, the valve opens or closes by a fixed amount each time a pulse is received from the computer. The computer can open the valve all the way by pulsing the open control input a certain number of times. The same number of pulses on the close control input will completely close the valve.

Figure 7-9.
Idle Air Control

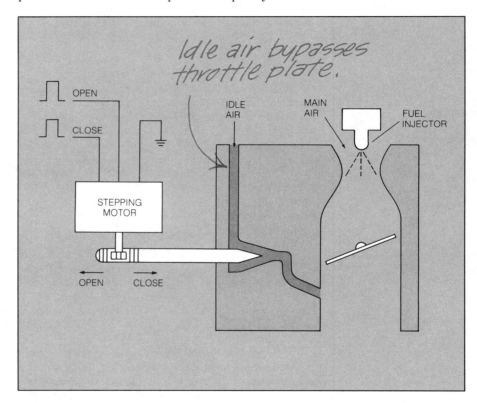

Idle speed is detected by the RPM sensor and the speed is adjusted to maintain the idle RPM constant. The computer receives digital on-off status inputs from several power take-off devices attached to the engine such as the air-conditioner clutch switch, park-neutral switch, and battery charge indicator to indicate the load that is applied to the engine during idle.

When the engine is not idling, the idle speed control valve is completely closed so that the throttle plate has total control of intake air. During periods of deceleration enleanment, the idle speed valve may be opened to provide extra air to increase the air/fuel ratio to reduce HC emissions.

EGR CONTROL

A second control subsystem of electronic engine control is the control of exhaust gas that is recirculated back to the intake manifold. Under normal operating conditions, engine cylinder temperatures can reach more than 3000°F. The higher the temperature, the more chance the exhaust will have NO_x emissions. A small amount of exhaust gas is introduced into the cylinder to replace normal intake air. This results in less air/fuel mixture entering the cylinder, lower combustion temperatures, and lower NO_x emissions.

The control mode selection logic determines when EGR is turned off or on. EGR is turned off during cranking, cold engine temperature (engine warm up), idling, and acceleration or other conditions demanding high torque.

The EGR control signal is determined as shown in *Figure 7-10* using inputs of RPM, coolant temperature and engine load (throttle position). The EGR signal can either control a valve opening which is detected by a valve position sensor, or it can be a duty cycle signal that meters EGR the same way as TBFI meters fuel. A valve positioning sensor is shown in *Figure 7-10*.

The engine controller also must determine when the EGR valve should be opened or closed. The EGR valve is closed during cranking, warmup, idling, acceleration, or heavy engine load.

**Figure 7-10.
EGR Control**

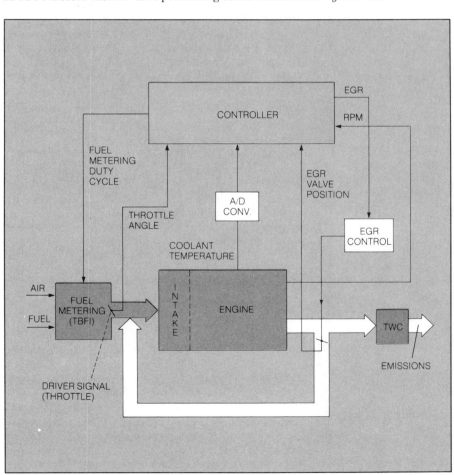

Some systems use an EGR valve which is controlled directly by manifold vacuum and is not under the control of the computer. To perform the open loop air/fuel ratio calculations, the computer must know how much EGR is being fed into the air intake. This is determined by using a sensor similar to the throttle position sensor that gives an electrical signal proportional to the amount of opening of the EGR valve.

ELECTRONIC SPARK CONTROL

The third control subsystem is electronic spark control. Electronic ignition has already been discussed; therefore, this discussion centers on digital system control of spark timing as would occur in a computerized control system.

The mode control selection logic in the computerized system's program selects the method of spark timing control. During engine starting, spark timing is controlled by the mechanical setting of the distributor. Once the engine is running, spark timing is turned over to the computer control system. This scheme ensures that the can will start regardless of whether the digital control system is working or not.

The controller uses look-up tables for RPM, coolant temperature, and MAP values to calculate the proper amount of timing advance to produce maximum engine torque under varying conditions.

The goal of electronic spark timing is to produce maximum engine torque by adjusting the advance of the ignition firing in relationship to TDC. The spark timing can be chosen to produce the best engine torque with input variables of engine RPM, engine coolant temperature, initial MAP (barometric pressure), and operating MAP.

Computing Spark Advance

The values from these tables are added, and the initial advance subtracted from their sum to produce a final value of spark advance.

The total spark advance is determined by adding together the contribution of each of the parameters affecting spark timing. Look-up tables are provided in the computer system to accomplish this. There is one for warm-up spark advance, SA_T, because more advance is required (or desirable) as the engine warms. This table lists SA_T vs temperature. There is one for a special spark advance, SA_S, to improve fuel economy during steady state conditions. This table lists SA_S vs RPM and load. And there is one for a spark advance adjustment due to barometric pressure, SA_P. Here, SA_P, is listed vs barometric pressure which is initial no-run MAP. All of the spark advances from these various look-up tables are added together and the initial mechanical advance is subtracted to determine the final spark advance SA_F. The control computations are indicated in *Figure 7-11a*.

**Figure 7-11.
Computerized Spark
Advance**

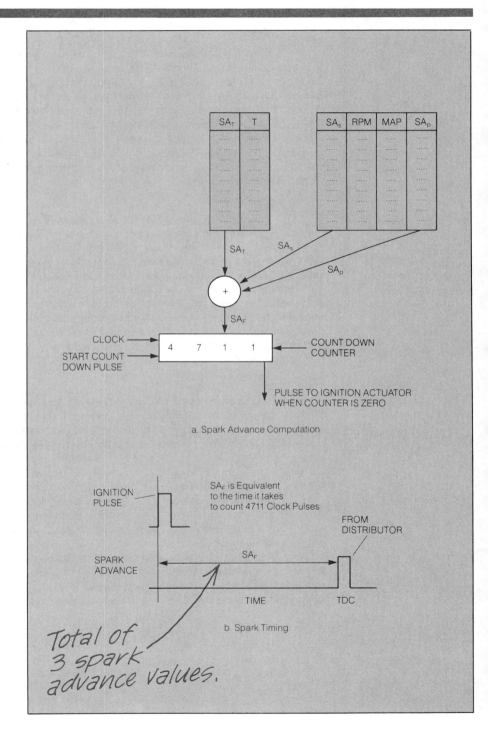

a. Spark Advance Computation

b. Spark Timing

Total of 3 spark advance values.

Spark Timing

The computer controls spark timing by counting a calculated number of clock pulses following a timing pulse from the crankshaft TDC position sensor. When the counter reaches zero, the computer sends a pulse to the ignition circuit to trigger the spark.

The computer receives a timing pulse from a position sensor which indicates crankshaft position for TDC as shown in *Figure 7-11b*. The computer loads a count-down counter (*Figure 7-11a*) with the computed spark advance number, SA_F, and the counter decrements that number until it reaches zero. At that time, the computer sends a pulse to the ignition actuator circuit which opens the ignition coil primary to generate a secondary voltage pulse to fire the spark plug. In some cases, the circuitry to open the primary of the coil may be in the computerized controller. The spark plug selection is performed mechanically by the distributor rotor and contacts as it has been done in the former all mechanical systems.

Secondary System

Some automotive systems perform spark advance as a secondary function of electronic fuel control.

Although it is possible for the electronic fuel control system to operate independently from the spark advance control system, electronic spark advance can be included as a secondary function of the fuel control system in certain cases because the electronic fuel control system has "excess capacity". As a secondary function in such a fuel control system based upon a TWC where air/fuel ratio is held at stoichiometry, the spark advance can be set to optimize engine performance (i.e., minimum advance for best torque or maximum fuel economy) because torque, fuel economy and emissions are dependent on spark advance. The required spark advance for this optimum performance can be experimentally determined at any operating condition during the engine mapping procedure.

Closed Loop System

Most spark advance control systems are open loop controllers like the one described above. Unfortunately, an open loop control system cannot adjust for mechanical changes or for changes in major external variables. A closed loop system which monitors such system changes and adjusts spark advance as needed would be much better. A complete system of this type will still need significant development because of the unavailability of economical sensors, but some work has been done on closed loop systems.

There have been a number of experimental closed loop electronic spark advance control systems. Some of these operate by sensing the occurrence of excessive knock which usually happens whenever the spark is advanced too much.

Engine knock is a condition where the air/fuel mixture in the cylinder does not burn normally. The pressure rise during this burning is so rapid compared to normal combustion that it is accompanied by an audible "knock" as though the engine were being struck by a hammer repeatedly as each cylinder "fires".

It is important to avoid knock, though some relatively low level may be acceptable for efficient operation. As mentioned, one concept for a closed loop system is based upon detecting the occurrence of knock. The spark is advanced until the maximum acceptable level of knock is reached. The closed loop control system using this scheme is often referred to as "knock limited spark advance".

One knock detection system that has been tried consists of an accelerometer mounted on the cylinder head. Such a detector produces an output voltage depending on the amount of engine vibration occurring in a certain frequency band. The "pinging" sound produced by the spark knock falls within this band. When the computerized controller receives a signal from the detector, it backs off (retards) the spark advance until the knocking stops and then starts increasing it again. This cycle is repeated constantly and makes sure that the spark advance is always maximum. Such a system can only indicate too much spark advance and has not yet achieved its full potential because it mistakes other engine noises for knock.

Another interesting concept for controlling spark advance is based upon instantaneous cylinder pressure during the power stroke. Cylinder pressure rises rapidly following ignition, reaches a maximum, then decreases. Research and experimentation have shown that keeping the crankshaft angle corresponding to peak cylinder pressure at about 15° after TDC leads to best fuel economy for certain engines regardless of external influences. Therefore, a closed loop spark advance control system based upon measurement of the crankshaft angle for maximum cylinder pressure could be built. The electronic closed loop control system would then adjust spark advance to maintain peak cylinder pressure at the optimum angle.

None of these systems for closed loop control of spark advance thus far have proved acceptable for widescale production. Hopefully, there will be a major breakthrough in closed loop spark control in the future.

INTEGRATED ENGINE CONTROL SYSTEM

Each control subsystem for fuel control, spark control and EGR has been discussed separately. However, as indicated in *Figures 6-5* and *6-6*, a fully integrated electronic engine control system can include these subsystems and provide additional functions. Usually the flexibility of the digital computerized system allows such expansion quite easily because the computer program can be changed to accomplish the expanded functions. Let's look at several of these.

Secondary Air Management

Secondary air management is used to improve performance of the catalytic converter. During engine warm-up, secondary air is routed to the exhaust manifold to speed warmup of the converter.

Secondary air management is used to improve the performance of the catalytic converter by providing extra oxygen rich air to either the converter itself or the exhaust manifold. The catalyst temperature must be above about 200°C to efficiently oxidize HC and CO and reduce NO_x. During engine warm up when the catalytic converter is cold, HC and CO are oxidized in the exhaust manifold by routing secondary air to the manifold. This creates extra heat to speed warm-up of the converter and EGO sensor which enables the fuel controller to go to the closed loop mode more quickly.

The converter can be damaged if too much heat is applied to it. This can occur if large amounts of HC and CO are oxidized in the manifold during heavy loads calling for fuel enrichment or during severe deceleration. In such cases, the secondary air is directed to the air cleaner where it has no effect on exhaust temperatures.

After warm up, the main use of secondary air is to provide an oxygen rich atmosphere in the second chamber of the three-way catalyst dual chamber converter system. In a dual chamber converter, the first chamber contains rhodium, palladium and platinum to reduce NO_x and to oxidize HC and CO. The second chamber contains only platinum and palladium. The extra oxygen from the secondary air improves the converter's ability to oxidize HC and CO in the second converter chamber.

The computer program for the control mode selection logic is modified to include the conditions for secondary air.

The computer controls secondary air by using two solenoid-operated valves that route air to the air cleaner, exhaust manifold, or directly to the converter.

The computer controls secondary air using two solenoid valves similar to the EGR valve. One valve switches air flow to the air cleaner or to the exhaust system. The other valve switches air flow to the exhaust manifold or to the converter. The air routing is based on engine coolant temperature and air/fuel ratio. The control system diagram is shown in *Figure 7-12*.

Evaporative Emissions Canister Purge

During engine off conditions, the fuel stored in the carburetor tends to evaporate into the atmosphere. To reduce these HC emissions, they are collected by a charcoal filter in a canister. The collected fuel is drained into the carburetor through a solenoid valve controlled by the computer. This is done during closed loop operation to reduce fuel calculation complications in the open loop mode.

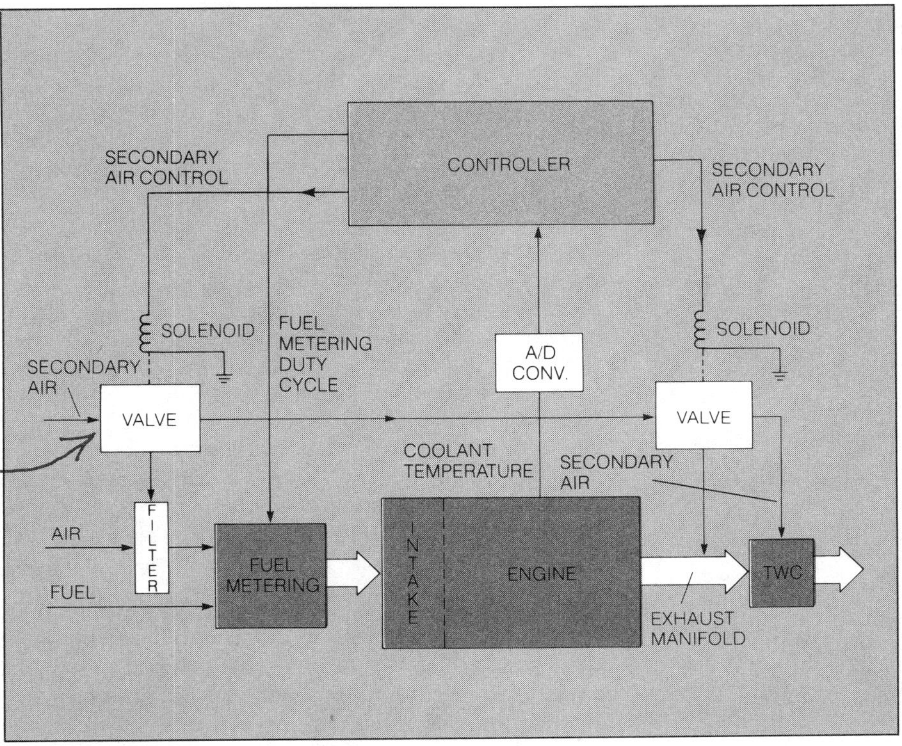

**Figure 7-12.
Secondary Air**

Secondary air either to intake, exhaust, or converter.

Torque Converter Lock-Up Control

Automatic transmissions use a hydraulic or fluid coupling to transmit engine power to the wheels. Because of slip, the fluid coupling is less efficient than the non-slip coupling of a pressure plate manual clutch used with a manual transmission. Thus, fuel economy is usually lower with an automatic transmission than with a standard transmission. This problem has been partially remedied by placing a clutch similar to a standard pressure-plate clutch inside the torque converter of the automatic transmission and engaging it during periods of steady cruise. This enables the automatic transmission to provide fuel economy near that of a manual transmission and still retains the automatic shifting convenience.

Here is an example of the ease of adding a function to the electronic engine control system. The torque converter locking clutch is activated by a lock up solenoid controlled by the engine control system computer. The computer determines when a period of steady cruise exists from throttle position and vehicle speed changes. It pulls in the locking clutch and keeps it engaged until it senses conditions that call for disengagement.

When the engine controller detects periods of steady cruise, it can energize a solenoid-operated lock-up clutch in the automatic transmission torque converter to increase fuel economy at cruise speeds.

Automatic System Adjustment

Under automatic system
adjustment of look-up
table values, the computer
compares the stored open
loop table values with the
operational closed loop values. If the values differ
greatly, the controller
stores new, closely-
matched, open loop values
in the look-up tables for
future use.

Another important feature of microcomputer engine control systems is their ability to be programmed to learn from their past experiences. Many control systems use this feature to enable the computer to learn new look-up table values for computing open-loop air/fuel ratio. While the computer is in the closed loop mode, the computer checks its open loop calculated air/fuel ratios and compares them with the closed loop average limit cycle values. If they are matched closely, nothing is learned and the open loop look-up tables are unchanged. If the difference is large, the control system controller corrects the look-up tables so the open loop control control values more closely match the closed loop values. This updated open loop look-up table is stored in memory (RAM) which is always powered directly by the car battery so that the new values are not lost while the ignition key is turned off. The next time the engine is started, the new look-up table values will be used in the open loop mode and will provide more accurate control of air/fuel ratio. This feature is very important because it allows the controller to adjust to long term changes in engine and fuel system conditions. This feature can be applied in individual subsystem control systems or in the fully integrated system. If not available initially, it may be added to the system by modifying the control program.

System Diagnosis

Another important feature of microcomputer engine control systems is their ability to diagnose failures in the control system and alert the operator. Sensor and actuator failures or misadjustments can be easily detected by the computer. For instance, the computer will detect a malfunctioning MAP sensor if the sensor's output goes above or below certain specified limits or fails to change for long periods of time. A prime example is the automatic adjustment system just discussed. If the open loop calculations consistently come up wrong, the engine control computer may determine that one of the many sensors used in the open-loop calculations has failed.

Abnormal responses from
sensors or actuators can
be detected by microcom-
puter engine control sys-
tems. The system can
switch to an alternate
means of engine control,
and alert the driver by
means of a dashboard
indicator.

If the computer detects the loss of a primary control sensor or actuator, it may choose to operate in a different mode until the problem is repaired. The operator is notified of a failure by blinking lights or some other indicator on the dashboard. Because of the flexibility of the microcomputer engine control system, additional diagnostic programs might be added due to different engine models that contain more or less sensors. Keeping the system totally integrated gives the microcomputer controller access to more sensor inputs so they can be checked.

SUMMARY OF CONTROL MODES

Now that a typical electronic engine control system has been discussed to show how the functions are accomplished and what inputs are used and the outputs that result, let's summarize what happens in an integrated system operating in the various modes.

Engine Crank (start)

Table 7-1 summarizes the engine operation in the engine crank (starting) mode. Primary control concern is for reliable engine start.

**Table 7-1.
Engine Crank**

1. Engine RPM at Cranking Speed
2. Engine Coolant at Low Temperature
3. A/F Ratio Low
4. Spark Timing Controlled Mechanically
5. EGR Off
6. Secondary Air to Exhaust Manifold
7. Fuel Economy Not Closely Controlled
8. Emissions Not Closely Controlled

Engine Warm Up

Table 7-2 summarizes the engine operations while the engine is warming up. The engine temperature is rising to its normal value. Primary control concern is for rapid and smooth engine warm up.

**Table 7-2.
Engine Warm-Up**

1. Engine RPM Above Cranking Speed at Command of Driver
2. Engine Coolant Temperature Rises to Minimum Threshold
3. A/F Ratio Low
4. Spark Timing Set by Controller
5. EGR Off
6. Secondary Air to Exhaust Manifold
7. Fuel Economy Not Closely Controlled
8. Emissions Not Closely Controlled

Open Loop Control

Table 7-3 summarizes the engine operations when the engine is being controlled with an open loop system. This is before the EGO sensor has reached the correct temperature for closed loop operation. Fuel economy and emissions are closely controlled.

**Table 7-3.
Open Loop Control**

1. Engine RPM at Command of Driver
2. Engine Temperature Above Warm-up Threshold
3. A/F Ratio Controlled by an Open Loop System to 14.7
   Regulated by the Mass of Air Entering the System
4. EGO Sensor Temperature Less Than a Minimum Threshold
5. Spark Timing Set by Controller
6. EGR Controlled
7. Secondary Air to Catalytic Converter
8. Fuel Economy Controlled
9. Emissions Controlled

Closed Loop Control

For the closest control of emissions and fuel economy under various driving conditions, the electronic engine control system is in a closed loop. *Table 7-4* summarizes the engine operation. Fuel economy and emissions are controlled very tightly.

Table 7-4.
Closed Loop Control

1. Engine RPM at Command of Driver
2. Engine Temperature at Normal (Above Warm-up Threshold)
3. Average A/F Ratio Controlled to 14.7 ± 0.05
4. EGO Sensor's Temperature Above Minimum Threshold Detected by a Sensor Output Voltage Indicating a Rich Mixture of Air and Fuel for a Minimum Amount of Time
5. System Returns to Open Loop if EGO Sensor Cools Below Minimum Threshold or Fails to Indicate Rich Mixture for Given Length of Time
6. EGR Controlled
7. Secondary Air to Catalytic Converter
8. Fuel Economy Tightly Controlled
9. Emissions Tightly Controlled

Hard Acceleration

When the engine must be accelerated quickly or if the engine is under heavy load, it is in a special mode summarized by *Table 7-5*. The engine controller is primarily concerned with providing maximum performance.

Table 7-5.
Hard Acceleration

1. Driver Asking for Sharp Increase in RPM or in Engine Power, Demanding Maximum Torque
2. Engine Temperature in Normal Range
3. A/F Ratio Rich Mixture
4. EGO Not in Loop
5. EGR Off
6. Secondary Air to Intake
7. Relatively Poor Fuel Economy
8. Relatively Poor Emissions Control

Deceleration and Idle

Slowing down, stopping and idling is another special mode. The engine operation is summarized in *Table 7-6*. The engine controller is primarily concerned with reducing excess emissions during deceleration and keeping idle fuel consumption at a minimum.

**Table 7-6.
Deceleration and Idle**

1. RPM Decreasing Rapidly Due to Driver Command or Else Held Constant at Idle
2. Engine Temperature in Normal Range
3. A/F Ratio Lean Mixture
4. Special Mode in Deceleration to Reduce Emissions
5. Special Mode in Idle to Keep RPM Constant at Idle as Load Varies Due to Air Conditioner, Automatic Transmission Engagement, Etc.
6. EGR On
7. Secondary Air to Intake
8. Good Fuel Economy During Deceleration
9. Poor Fuel Economy During Idle, but Fuel Consumption Kept to Minimum Possible

SUMMARY

In this chapter, we have discussed a typical digital electronic automotive engine control system based on a microcomputer which is used to improve automotive emissions and fuel economy. We discussed electronic fuel control, electronic spark timing, exhaust gas recirculation, and other electronically controlled subsystems which improve total system performance. These systems represent what is being done today with electronic engine controls. The details change from manufacturer to manufacturer, but they are all approaching the problem with basically the same point of view—to satisfy the government's emissions and fuel economy requirements by using cost effective, versatile, electronic microcomputer control systems.

Chapter 7 Quiz

1. A binary valued signal can be how many different voltages?
   a. 10
   b. 16
   c. 2
   d. infinitely many

2. An analog signal can be how many different voltages?
   a. 10
   b. 16
   c. 2
   d. infinitely many

3. Which device converts analog signals to digital numbers?
   a. A/D converter
   b. D/A converter
   c. low pass filter
   d. binary counter

4. Automotive sensors use what type of signals?
   a. analog
   b. digital
   c. pulse
   d. all of the above

5. The controller discussed in this chapter is what type of controller?
   a. analog
   b. digital
   c. both analog and digital

6. A low air/fuel ratio is what type of fuel mixture?
   a. lean
   b. rich
   c. poor
   d. fat

7. Open and closed loop fuel control systems control air/fuel ratio near which of the following?
   a. rich
   b. lean
   c. stoichiometry
   d. rich and lean

8. Acceleration enrichment is used for what purpose?
   a. reduce fuel consumption
   b. reduce exhaust emissions
   c. provide maximum torque
   d. provide minimum fuel economy

9. Idle speed control is used for what reason?
   a. maximum idle speed
   b. minimum idle fuel consumption
   c. deceleration enleanment
   d. maximum torque

10. Microcomputer engine controls allow which of the following features?
    a. multivariable control
    b. self adjustment
    c. malfunction indication
    d. all of the above

Vehicle Motion Control

ABOUT THIS CHAPTER

Electronic controls can automate some driver functions that were previously performed manually.

Electronic control in automobiles is making it much easier for the driver. With electronic engine control, the driver merely selects throttle setting, the appropriate brake pressure and does the steering, and the engine controller and automatic transmission controls the rest. The engine controller optimizes fuel economy and exhaust gas emissions; the automatic transmission shifts at optimum points in the drive cycle without driver attention.

In this chapter, additional electronic control systems for vehicle motion that add to the convenience and safety of the driver and passengers will be discussed. The first system, commonly referred to as cruise control, basically maintains the speed (velocity) of the vehicle constant.

TYPICAL CRUISE CONTROL SYSTEM

A cruise control is a closed loop system that uses feedback of vehicle speed to adjust throttle position.

Figure 8-1 shows a block diagram of a cruise control system. The control electronics has two inputs, the command speed signal that indicates the desired speed and the feedback speed signal that indicates the actual vehicle speed. The control electronics detects the difference between the two inputs (the error) and produces a throttle control signal which is sent to the throttle actuator. The throttle actuator sets the engine throttle position which alters

Figure 8-1.
Block Diagram of Cruise Controller

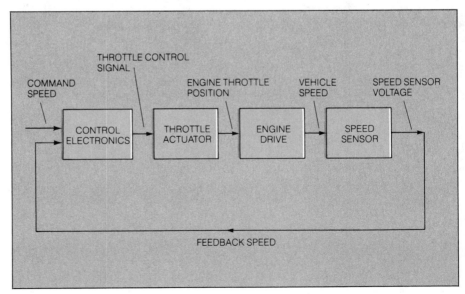

the engine speed to correct for the vehicle speed error detected by the control electronics. The vehicle speed is detected by the speed sensor and converted to an electrical voltage proportional to vehicle speed. The control system is operating closed loop because the speed signal is fed back to the control electronics to be compared to the command signal.

How Cruise Controllers Work

Cruise controllers adjust speed by increasing or decreasing throttle angle in response to feedback that indicates whether actual speed is above or below the desired speed. Time lags in vehicle response must be taken into account in controller design.

Cruise controllers regulate vehicle speed by adjusting the engine throttle angle to increase or decrease the engine drive force depending on whether the speed is below or above the command value. The controller has to take into account the time lag between its newly commanded drive force and the resulting final speed. The vehicle speed can become unstable and oscillate (vary up and down) if the controller tries to correct speed errors too quickly.

In Chapter 2, we talked briefly about proportional controllers. One of the major drawbacks of using proportional control alone is the size of the steady state error allowed by proportional control systems. The final error depends on the proportional gain constant. To reduce the final error to a very small difference requires a large control effort or gain.

Most speed controllers use proportional-integral (PI) control. The control signal is the sum of two signals, one that is proportional to the error signal and one whose ramp rate is proportional to the error signal.

Most vehicle speed controllers use a type of control electronics known as proportional-integral control, or PI control. This type of controller is shown in *Figure 8-2*. The control signal is actually the sum of two other signals. The proportional gain block K_p provides a control signal which is proportional to the error signal, e. The other block, the integrator block, is different. It provides a signal that slopes or ramps up or down. The slope or rate at which the signal ramps up or down is proportional to the error signal, e. The gains K_p and K_I are chosen so that the system has quick response, high accuracy, and no instability or oscillations.

Integral control is used in addition to proportional control to help drive the steady state error to zero. It does this by effectively adding up the error as time goes by. The integral control block always causes the final error to tend towards zero. The time required to drive the error to zero is determined by the integral control gain.

Performance Curves

The performance of such PI systems can be described by plotting the vehicle speed (velocity) response of the system of *Figure 8-2* against time when a small change is made in the commanded speed.

Figure 8-2.
Block Diagram of Cruise
Controller Using PI
Control

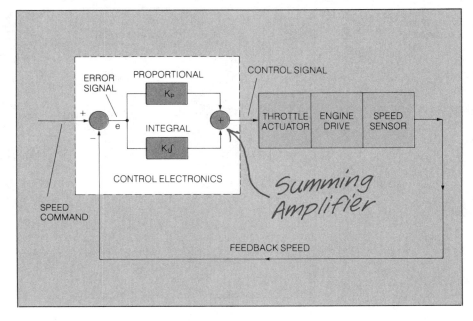

Speed Response Curves

When a new speed is requested, the time required for the vehicle to reach that speed is affected by the control system's damping coefficient.

Figure 8-3 shows the plot of how the vehicle speed varies with time when a new command speed is requested. The system cannot respond in an instant. It takes time, and as shown in *Figure 8-3*, the time that it takes is specified by the time constant of the system measured to the point where the speed is 63.2% of what its final value will be. The time constant varies with the parameters of the complete system. However, a major controlling parameter is the damping coefficient. The curves of *Figure 8-3* are plots for systems with different damping coefficients.

Systems that have an over damped response are sluggish. Systems that are under damped provide quick response, but oscillate. Critically and optimally damped systems come closer to an ideal response.

Curve 1 is the speed response of a system that is over damped. The speed of the system rises sluggishly and takes a long time to reach the command velocity. It has a time constant determined by point A. Curve 4 is the speed reponse of a system that is under damped. The speed curve rises quickly, but overshoots the commanded speed and then oscillates around the commanded speed until it finally settles down. The frequency of the oscillations is called the system's natural frequency. The time constant is determined by point D.

Curves 2 and 3 are systems that have damping coefficients that make the system a critically damped or an optimally damped system. The critically damped system rises smoothly to the command speed with no overshoot and in a minimum amount of time. The optimally damped system rises quickly and overshoots a little, but settles quickly to the commanded velocity. This system is called optimally damped because the vehicle speed follows the command speed more closely than for any other of the systems.

**Figure 8-3.
System Speed
Performance**

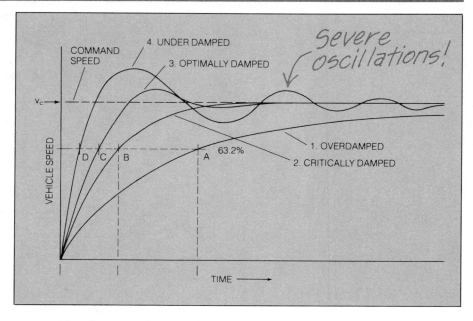

Designers usually strive for an optimally damped control system, but engine power limits and other practical limitations may force the designer to choose a different response.

Usually a control system designer attempts to juggle the proportional and integral control gains so that the system is optimally damped. However, because of system characteristics, it is impossible, impractical, or inefficient in many cases to achieve the optimal time response, so another response is chosen. The control system should make the engine drive force react quickly and accurately to the command speed, but shouldn't use excessive effort (overtax the engine) in the process. Therefore, the system designer chooses the control electronics based on the following system qualities:

1. Quick response
2. Relative stability
3. Small steady state error
4. Optimize the control effort required

Frequency Response Curves

Another way of describing the system performance is to show the response of the system to command signals that are changing in frequency. Recall that frequency is the number of cycles of signal change in one second. By using performance curves like these, the control system designer is able to determine the value of the control system gains K_p and K_I that are required.

When the command speed is varied at a set rate or frequency, the response of the system to this change can be displayed on a log-arithmic graph.

Figure 8-4a shows system gain, K_S, frequency response curves of a closed loop system with different damping coefficients (see Chapter 2). These curves assume that the command speed input signal is being changed in a sine-wave fashion. Again, as with the speed response curves, the command speed inputs are changing small amounts around a normal operating speed within the control range of the system, as shown in *Figure 8-4c*.

**Figure 8-4.
Vehicle Motion Closed
Loop Frequency
Response Plot**

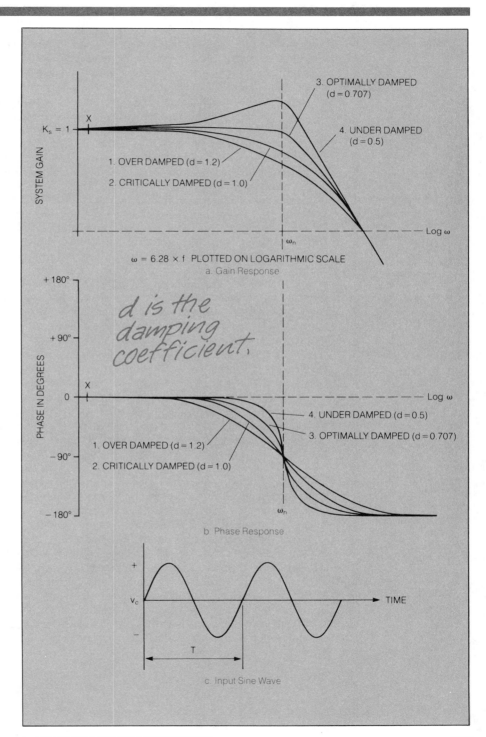

3. OPTIMALLY DAMPED
(d = 0.707)

4. UNDER DAMPED
(d = 0.5)

$K_s = 1$

X

SYSTEM GAIN

1. OVER DAMPED (d = 1.2)

2. CRITICALLY DAMPED (d = 1.0)

Log ω

ω_n

ω = 6.28 × f PLOTTED ON LOGARITHMIC SCALE
a. Gain Response

d is the damping coefficient.

+ 180°

+ 90°

PHASE IN DEGREES

X

0

Log ω

4. UNDER DAMPED (d = 0.5)

3. OPTIMALLY DAMPED (d = 0.707)

1. OVER DAMPED (d = 1.2)

2. CRITICALLY DAMPED (d = 1.0)

- 90°

- 180°

ω_n

b. Phase Response

+

V_c

TIME

−

T

c. Input Sine Wave

The command speed is varied plus or minus around V_c in a sine wave fashion. When the frequency of the sine wave is a low frequency, the amplitude of the system gain is 1 and the operating point is at point X on the gain curves of *Figure 8-4a* and the phase curves of *Figure 8-4b*. The frequency, f, is plotted on the frequency axis as 6.28 times f because it is easier to use in engineering calculations. The frequency scale is a logarithmic scale rather than linear so that it is compressed and tens of cycles and millions of cycles can be on the same graph. An example of a low frequency for such a system might be where the period of the sine wave (T in *Figure 8-4c*) is 10 seconds. If T = 10 seconds, then the frequency of the input command speed signal is one cycle in 10 seconds, or 0.1 cycle per second.

As the frequency (rate of change) of the command speed increases, system gain remains relatively constant to a given point, then changes rapidly *(Figure 8-4a)*.

As the input frequency is increased the system gain stays constant for a certain band of frequencies and then starts changing. Note that the system with different damping coefficients have different frequency response curves. For underdamped systems, the system gain actually increases over the low-frequency gain at point X. This increase corresponds to the system response shown in curve 4 of *Figure 8-3*. When the speed command is abruptly changed the system overshoots and oscillates around the command speed setting. A similar correspondence between the curves of *Figure 8-4* exists for the other curves of *Figure 8-3*. The sluggish response of a system indicated by curve 1 of *Figure 8-3* corresponds to an overdamped system frequency response of *Figure 8-4a* where the system gain is not as good at higher frequencies as even the critically damped system.

To understand what happens to system response as the input signal frequency is increased, follow the optimally damped curve 3 in *Figure 8-4a* from point X. The gain, K_s, of the system remains constant until an input frequency of f_n is reached. This is represented by ω_n ($\omega_n = 6.28 \times f_n$). Up to this point the system response is the same as the lower frequency point X. As the input command speed signal frequency is increased beyond point ω_n, the system gain K_s starts to decrease from its low frequency value and continues to decrease as frequency is increased.

Once the frequency increases past a certain point, none of the systems can respond accurately to the rate of change. The system output, due to physical limitations, cannot keep up with the system input.

What the curves of *Figure 8-4a* are saying is that over a band of frequencies from very low to a frequency f_n (represented by ω_n), the system response remains the same, but after ω_n, the system response gets poorer because the system gain is decreasing. Said another way, the system output can no longer keep up with the changes being made at the input.

This is further substantiated by the curves of *Figure 8-4b*, which are called phase response curves. These curves tell us how well the system output is following the input. To illustrate, suppose an engine is turning the input shaft of a system that has a fluid coupling between the input shaft and the output shaft. As the input shaft is turned at a faster speed or as the output shaft is loaded with a greater load, the output shaft will begin to lag the input shaft because of slip in the fluid coupling. Phase can be thought of as the number of degrees the output shaft lags the input shaft in one revolution.

With this thought in mind, curve 3 of *Figure 8-4b* shows that as the frequency of the input signal is increased, the response of the system begins to lag the input (minus degrees) because it can't keep up. When the input frequency is f_n (represented by ω_n), then the output is 90 degrees lagging.

CLOSED LOOP RESPONSE

The gain of a closed loop speed controller can be calculated, using formulas that take integral gain and proportional gain into consideration.

The loop gain K_s of the system depends on K_p and K_I of *Figure 8-2*. In order for the electronic controls to be designed properly, K_p and K_I must be known. The integral gain constant, K_I, can be determined from:

$$K_I = \omega_n^2 M$$

where M is the mass of the vehicle that is being controlled.

The proportional gain constant, K_p, can be determined from:

$$K_p = (2d\, \omega_n\, M) - C$$

where d is the damping coefficient, M is the mass of the vehicle that is being controlled, and C is a friction factor that is determined experimentally by measuring the engine drive force that is lost because of mechanical friction of turning parts, resistance of the air and resistance of the tires against the road.

Design Choices

Total system response is affected by many factors. Vehicle mass, friction, engine power, and the desired damping coefficient all must be entered into calculations performed by the system designer.

For any automotive electronic speed control, the system designer makes several choices. The decision process might go something like this: First, the mass of the vehicle, the engine power, and the friction factor C (composed of a number of friction components) are determined. Second, ω_n of the system is chosen. The input signal frequency range to which the controller must respond is based on a study of how quickly a human driver can detect and correct for vehicle speed changes on a hilly road surface. It is assumed that the cruise control system must respond at least as fast as the human. Humans respond in tens of milliseconds to tenths of seconds. From ω_n, K_I is set for a given M.

Third, the damping coefficient for the kind of response desired is chosen from the values as given in *Figure 8-4a* and *Figure 8-4b* using *Figure 8-3* as a guide. Knowing d, ω_n, M and C; K_p is determined.

The parameter ω_n and the control gains also are affected by the sensors and actuators of the system. Therefore, the system designer juggles the proportional and integral gains until the total response of the system is satisfactory.

SENSORS AND ACTUATORS

Before looking at actual cruise control electronic circuitry, let's examine the speed sensor and throttle actuator that might be used in the system.

Speed Sensor

Speed sensors are normally driveshaft-coupled, and can provide digital output (pulses) or analog outputs that are converted into digital signals.

The vehicle speed sensor is normally coupled to the vehicle driveshaft since its rotational speed is directly proportional to the vehicle speed. As discussed in Chapter 5 under RPM sensors, the vehicle speed sensor may provide an analog voltage output that is proportional to speed or it may give a digital output. The digital output can be derived by counting a given number of pulses in a particular time or it could be the output of an analog to digital converter derived from the analog voltage speed sensor.

In either case, there is an important point in the selection of the speed sensor as shown in *Figure 8-5*. The sensor frequency response is plotted on a frequency scale like *Figure 8-4* along with the open-loop system frequency response. The sensor should be chosen so that its frequency response is much higher than the open-loop system frequency response so that it is not a major contributor to the system frequency response.

**Figure 8-5.
Sensor Frequency
Response**

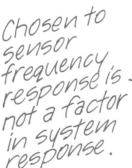

Chosen to sensor frequency response is not a factor in system response.

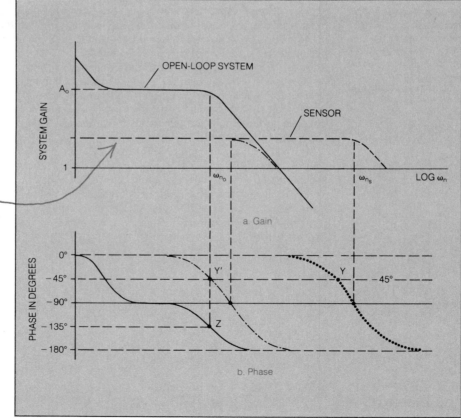

This can be illustrated as follows: Point Y on the phase curve of *Figure 8-5b* is the point where the sensor output lags the sensor input by 45°. The sensor gain at this point is still constant at its low frequency value *(Figure 8-5a)*. Point Z is the point on the phase curve of the open-loop system where the output lags the input by $-135°$. The open-loop system gain, A_o, is still much greater than one at this point on the frequency scale.

The rule in a closed-loop system to make it stable is that the open-loop system phase shift must not be equal to $-180°$ before the open-loop system gain has decreased to below one; otherwise, the system will oscillate.

When the sensor frequency response is such that point Y is far removed from point Z, then the open-loop system gain will be less than one where the phase shift of the sensor adds to the system phase shift. However, suppose the frequency response of the sensor were such that point Y were really at point Y'. Then the $-45°$ phase lag of the sensor would add to the $-135°$ phase lag of the open-loop system to produce a $-180°$ phase lag while the open-loop system gain is well above one. Under these conditions, the closed-loop system would not be stable and would oscillate. With the open-loop system gain greater than one and the phase shift equal to $-180°$, the vehicle speed signal is actually making the error signal greater rather than reducing it to zero.

In other words, if the vehicle speed is below the command input, the error signal will actually cause the vehicle to slow down rather than speed up. Eventually the time lag will be overcome and the large error signal will cause the vehicle to accelerate rapidly to reach the commanded speed. The phase lag will now let the speed overshoot the command input and the vehicle speed will be too fast.

Throttle Actuator

An actuator important to the cruise control system that was not covered in Chapter 5 is the throttle actuator. In all our discussions in earlier chapters, we have assumed the throttle was being actuated by the vehicle driver who is outside the electronic control system. With the cruise control, throttle actuation is a part of the electronic control system. Most throttle actuators use a type of pneumatic piston arrangement which is driven from the intake manifold vacuum. *Figure 8-6* shows a typical throttle actuator. The piston connecting rod is attached to the throttle lever. If there is no force applied by the piston, a spring pulls the throttle closed. When an actuator input signal energizes the electromagnet in the control solenoid, the pressure control valve is pulled down and changes the actuator cylinder pressure by providing a path to manifold pressure. Manifold pressure is lower than atmospheric pressure so the actuator cylinder pressure quickly drops causing the piston to pull against the throttle lever to open the throttle. The force exerted by the piston is varied by changing the average pressure in the cylinder chamber. This is done by rapidly switching the pressure control valve

---

Sensors must be chosen so that the sensors' frequency response does not degrade the frequency response of the total system. The sensors' frequency response should be much higher to ensure a stable system that will not oscillate.

Throttle actuators use manifold vacuum to pull a piston that is mechanically linked to the throttle. The amount of vacuum provided is controlled by a solenoid valve that is turned on and off rapidly.

A switching, duty cycle type of signal is applied to the solenoid coil. By varying the duty cycle, the amount of vacuum, and corresponding throttle angle, is varied.

between the outside air port which provides atmospheric pressure and the manifold pressure port, which is close to a vacuum. The actuator control signal, V_c, is a duty cycle type of signal like that discussed for the fuel injector actuator. A high V_c signal energizes the electromagnet; a low V_c signal de-energizes the electromagnet. Switching back and forth between the two pressure sources causes the average pressure in the chamber to be somewhere between the low manifold pressure and outside atmospheric pressure. The average pressure inside the chamber (and consequently the piston force) is proportional to the duty cycle of the valve control signal V_c.

**Figure 8-6.
Vacuum Operated
Throttle Actuator**

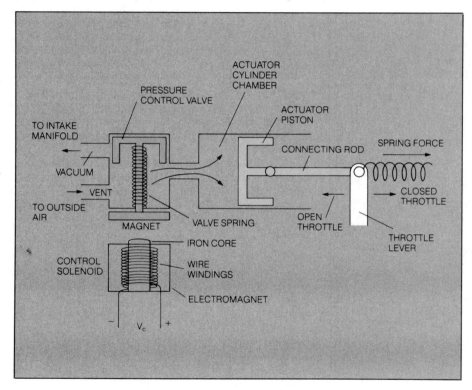

This type of duty cycle controlled throttle actuator is ideally suited for use in digital control systems. If used in an analog control system, the analog control signal must first be converted to a duty cycle control signal. The same frequency response considerations apply to the throttle actuator as to the speed sensor. In fact, with both in the control system closed loop, each contributes to the total system phase shift and gain.

CRUISE CONTROL ELECTRONICS

In an analog cruise control system, an error amplifier compares actual speed and desired (command) speed. The error signal output is fed to a proportional amplifier and an integral amplifier. The resultant outputs are combined by a summing amplifier.

The electronics for a cruise control system that is basically analog is shown in *Figure 8-7*. Notice that the system uses four operational amplifiers (op-amps) and each amplifier is used for a specific purpose. Op-amp 1 is used as an error amplifier. The output of op-amp 1 is proportional to the difference between the command speed and the actual speed. The error signal is then used as an input to op-amp 2 and op-amp 3. Op-amp 2 is a proportional amplifier with a gain of $K_p = -R_2/R_1$. Notice that R_1 is variable so that the proportional gain can be adjusted. Op-amp 3 is an integral amplifier or integrator with a gain of $K_I = -1/(R_3 C_1)$. R_3 is variable to permit adjustment of the gain. The op-amp causes a current to flow into capacitor C_1 which is equal to the current into R_3. The voltage across R_3 is the error amplifier output voltage, V_e. The current in R_3 is found from Ohm's law to be;

$$I = \frac{V_e}{R_3}$$

**Figure 8-7.
Cruise Control
Electronics**

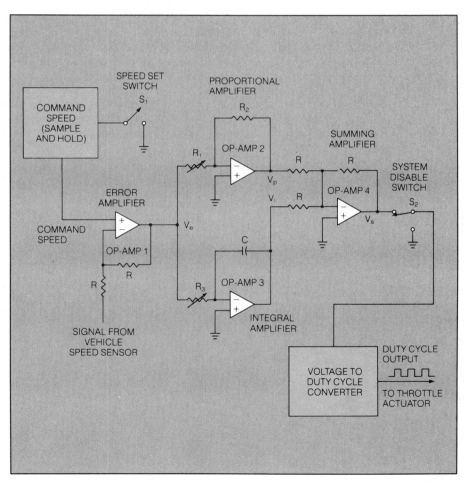

which is identical to the current into the capacitor. If the error signal V_e is constant, the current I will be constant and the voltage across the capacitor will steadily change at a rate proportional to the current flow.

The output of the integral amplifier, V_I, slopes up or down depending on whether V_e is above or below zero volts. The voltage V_I is steady or unchanging only when the error is exactly zero. That is why the integral gain block can reduce the system's steady state error to zero. Even a small error causes V_I to change to correct for the error.

The outputs of the proportional and integral amplifiers are added together using a summing amplifier, op-amp 4. The summing amplifier adds voltages V_p and V_I and inverts the resulting sum. The inversion is necessary because both the proportional and integral amplifiers invert their input signals while providing amplification. Inverting the sum restores the correct sense or polarity to the control signal.

The summing amplifier, op-amp 4, produces an analog voltage, V_s, which must be converted to a duty cycle signal before it can drive the throttle actuator. A voltage to duty cycle converter is used whose output directly drives the throttle actuator solenoid.

Figure 8-7 shows two switches, S_1 and S_2. Switch S_1 is operated by the driver to set the desired speed. It signals the sample and hold electronics to sample the current vehicle speed and remember it. The sample and hold circuit is shown in *Figure 8-8*. The voltage, V_1, representing the vehicle speed at which the driver wishes to set the cruise controller, is sampled and it charges a capacitor C. A very high input impedance amplifier detects the voltage on the capacitor without causing the charge on the capacitor to "leak" off. It outputs a voltage, V_s, proportional to the command speed to the error amplifier.

Because the output of the summing amplifier is an analog signal, it must be converted into a duty cycle signal to pulse the throttle actuator.

Figure 8-8.
Sample and Hold

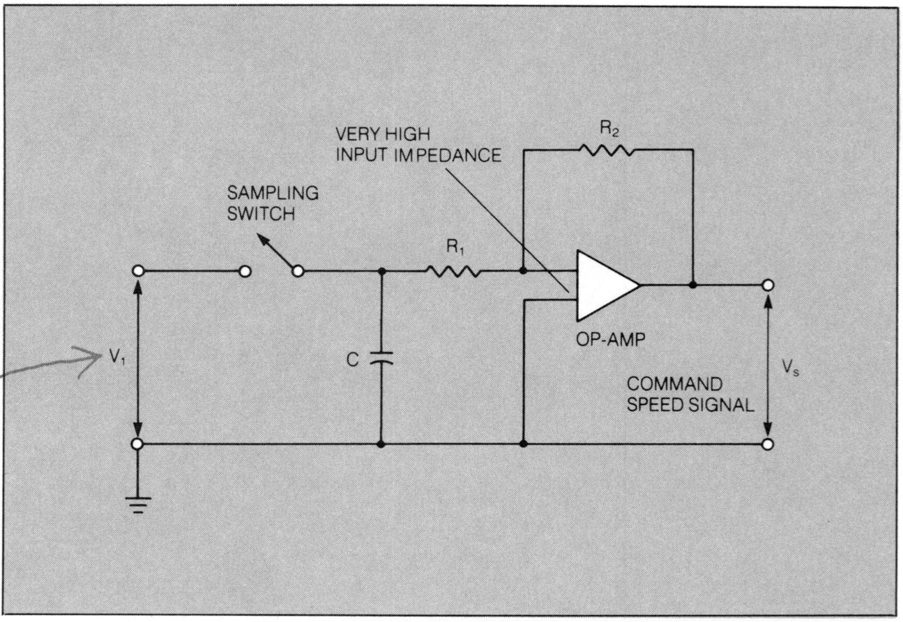

Sensor voltage (V_1) represents vehicle speed.

If the error detection circuit is digital, then the input voltage representing the command speed signal is input to an A/D converter and the digital number output is stored in latches. This number does not change or drift until a new command speed signal is inputted to the system.

Switch S_2 is used to disable the speed controller by interrupting the control signal to the throttle actuator. Switch S_2 disables the system whenever the ignition is turned off, the controller is turned off, or when the brake pedal is pressed. It turns on when the driver presses the speed set switch S_1.

For safety reasons, the brake turn off is performed two ways. As just mentioned, pressing the brake pedal turns off or disables the electronic control. The brake pedal also mechanically opens a separate valve which is located in a hose connected to the throttle actuator cylinder. When the valve is opened by depression of the brake pedal, it allows outside air to flow into the throttle actuator cylinder so the throttle plate instantly snaps closed. The valve is shut off whenever the brake pedal is in its inactive position. This ensures a fast and complete shut down of the speed control system whenever the driver presses the brake pedal.

Digital Speed Control

Digital speed controllers store command speed sensor pulses directly, and convert these values to binary codes. The codes are compared mathematically to produce the error signal used by the proportional gain and integrator gain logic.

As was mentioned earlier, the speed controller can be implemented using digital logic instead of analog circuitry. *Figure 8-9* shows a block diagram. When digital circuitry is used, the input speed command and the actual vehicle speed are stored directly as numbers. The actual speed signal is sampled periodically while the command speed signal is sampled only when the driver sets a new speed. The pulses coming from the speed sensor are counted in a given time period and converted to a digital number code representing the speed. The digital numbers are compared (really subtracted from each other to get a difference) and a number representing the error is operated on over two paths just as in the analog controller. The one path (a look-up table) outputs a number that represents the proportional gain for the system. The other path integrates the error numbers over time and outputs a number that represents the integral gain. An adder sums these numbers and a duty cycle generator provides the timed ON-OFF pulses to the throttle actuator.

Digital speed controllers have high stability in comparison to analog controllers. VSLI circuitry used in digital controllers provide increased programming flexibility.

A significant advantage that the digital circuitry has over analog circuitry is that the data or information contained in the system in digital codes does not change with severe temperature and humidity changes. Therefore, digital systems have much more stability over these extremes.

The digital controller can be a single integrated circuit contained in one package due to the advances made in LSI and VLSI technology, or the system can be accomplished by programming the microcomputer that is providing the electronic engine control. This demonstrates another distinct advantage of digital systems—additional functions can be added to a programmable digital system just by changing the program. This saves expensive control hardware.

Figure 8-9.
Digital Vehicle Speed
Controller

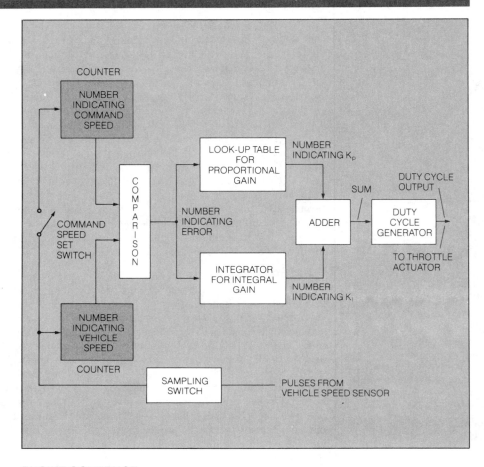

ENGINE GOVERNOR

Engine governors limit
maximum engine speeds
to prevent damage to en-
gine parts.

Another application of a speed controller is the engine governor. An engine governor controls and limits engine speed below some maximum limit without regard to vehicle speed. An engine governor is used mostly on heavy duty trucks and construction vehicles to prevent engine damage due to running them at too high a speed. Excessively high engine speed can damage engine bearings and cause severe strain on certain engine parts. Mechanical governors have been used in the past to control and limit engine speed.

The electronic engine governor controls and limits engine speed by adjusting the engine throttle (for an SI engine) with an actuator similar to the cruise control throttle actuator discussed earlier. Engine RPM is measured by the same type sensor as used to measure the vehicle speed; however, in this case, it may be mounted on the distributor shaft or the crankshaft rather than the driveshaft.

The engine governor control electronics closely matches that of the cruise control system of *Figure 8-7* except for some significant differences. One of these is that the engine speed may be allowed to vary through a range below a certain maximum limit. When the maximum limit is reached, the engine speed is maintained constant at or below the maximum limit. Another difference is that now only the engine speed is being controlled; therefore, the mass M and the friction C will be much less than for the complete vehicle involved in the cruise control system.

Anticipation

Engine governor control electronics utilize an additional signal called lead term. The lead term signal is used to predict future RPM by monitoring the rate of change in engine speed.

Since the engine speed is allowed to vary below the maximum limit, the governor must be able to detect how fast the engine speed is accelerating so the speed doesn't overshoot the limit before the governor control takes effect. To understand how this is done, let's examine the block diagram of *Figure 8-10*. The same proportional gain, K_p, and integral gain, K_I, are included in the engine governor block diagram as for the cruise control. However, now there is an additional block with a gain of K_d. The extra control block adds an extra signal to the PI control signal to take control of the throttle a little bit before the engine speed rises past the preset limit. The new control block is called a lead term because it anticipates or predicts engine speed at some time in the future. It makes this prediction based on the *rate of change* of the engine RPM. If the engine RPM is rising rapidly, the lead term control block adds a signal which takes control earlier than if engine RPM is rising slowly. This allows enough time for the system to respond so that the throttle angle is reduced before engine speed becomes excessive.

Figure 8-10.
Block Diagram of Engine Governor Control System

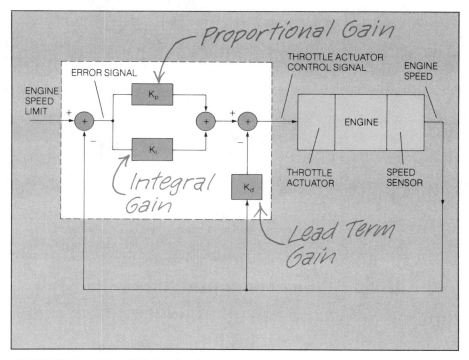

Lead Term Frequency Response

Lead term action purposely leads the system response to successfully predict the proper throttle angle.

Figure 8-11 shows the frequency response and phase shift for the lead term. The frequency response of *Figures 8-4* and *8-5* are for systems that are low pass filters. Such systems respond well to signals at low frequency, but have little response at high frequency. The lead term does just the opposite. It has a very low gain for low frequencies and has an increasing gain for increasing frequencies. The lead term gain becomes greater than 1 at frequency $\omega_3 = 1/K_d$. The $+90°$ phase shift indicates that the lead term action is ahead of (leads) the system response in order to predict the needed throttle angle as discussed above.

Figure 8-11.
Frequency Response of
Lead Term Control Block

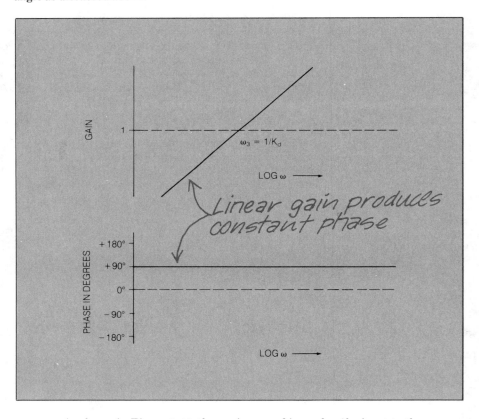

As shown in *Figure 8-10*, the engine speed is used as the input to the lead term. When the engine speed increases rapidly, the lead term reduces the engine throttle position in proportion to the rate of engine speed rise. This makes the engine feel a little bit sluggish because the effect of a quick accelerator depression, which normally produces a rapid increase in engine speed and fast acceleration, is reduced by the lead term. The amount of sluggishness is determined by the lead term gain K_d. A large K_d causes a very sluggish response while a small K_d allows a crisper response.

Governor Control Electronics

An analog governor control circuit uses a lead term op/amp to provide a gain constant with phase shift. Since the lead term amp inverts its signal, a second op amp corrects the inversion.

The control circuit for the engine speed governor is almost identical to the vehicle speed control schematic of *Figure 8-7*. The one difference is the inclusion of the lead-term circuitry. *Figure 8-12* shows the additional electronics for a lead term and how it fits into the circuitry of *Figure 8-7*. Another input is added to the summing amplifier to accept the lead-term signal in addition to the proportional and integral signals. The lead amplifier provides a gain constant, $K_d = C_1 R_1$. The lead amplifier is an inverting amplifier so another inverting amplifier is placed immediately before the lead circuit to correct the sign change.

This is the analog circuit solution. If the control is accomplished digitally, a program for the integrated control system that differentiates the signal must be added, or a separate digital circuit which is called a differentiator must be substituted in the system to accomplish what the lead circuitry does in the analog system.

Figure 8-12.
Lead Term Circuit for
Governor Control

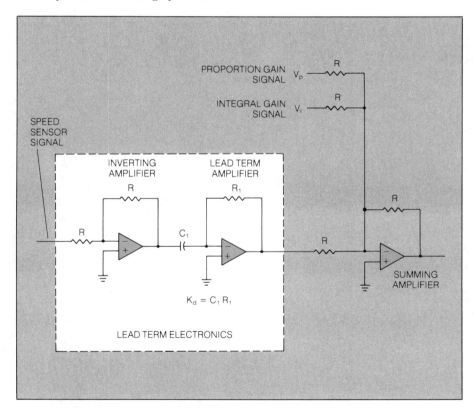

TIRE–SLIP CONTROLLER

In the cruise control system, all of the drive force from the engine to the drive train was considered to be transmitted to the wheels and tires to move the vehicle. However, in order to do this, the tire on the wheel must not slip on the running surface. If it does, a great deal of the driving force is lost.

Electronics has been used to develop a controller to detect tire slip and reduce engine torque to stop the slipping. Let's look at tire slip in a little more detail.

Figure 8-13 is a diagram of a tire running on a surface. The torque produced by the engine is converted to a force F which drives the vehicle in the direction shown. The force F is developed between the tire and the running surface because of static friction. As long as F is developed, the vehicle moves forward just as much as the wheel turns on the surface. However, if the force F can no longer be maintained by the friction between the tire and the surface, the tire slips on the surface and the vehicle forward motion is much less than the turning of the wheel. The wheel spins on the surface. The wheel speed is much greater than vehicle speed.

Tire-slip controllers detect tire slip by monitoring the speed relation between the driving wheels and the vehicle speed.

**Figure 8-13.
Tire Running on a
Surface**

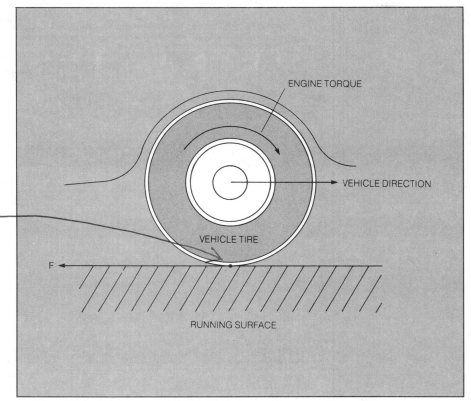

Slip occurs due to loss of friction.

Controller Electronics

Tire-slip controllers use a
predetermined voltage,
V_{set}, which represents the
maximum possible wheel
speed acceleration with no
wheel slip.

Figure 8-14 is a circuit of a controller that detects wheel slip by detecting an unusually high rate of change of wheel speed. The wheel speed sensor signal is first inverted by op-amp 1. Op-amp 2 is a lead term amplifier which is almost identical to the lead term amplifier discussed in the previous section on engine governors. The lead term amplifier produces a voltage output which is proportional to the rate of change of the input signal (the wheel speed signal). An increasing wheel speed produces a positive V_L at the lead amplifier output. A decreasing wheel speed produces a negative V_L. Op-amp 3 is a voltage comparator which has a low output voltage level if the lead amplifier output V_L is greater than V_{set}. V_{set} is chosen during design so that the maximum possible normal wheel speed acceleration causes V_L to be slightly below V_{set}. If the wheel acceleration is great enough to cause the lead amplifier output V_L to rise above V_{set}, the comparator output voltage level will go low. Under normal circumstances, V_L is less than V_{set} and the comparator output V_c is at a high level. A high V_c turns the transistor switch ON and the output voltage, V_o, is at a low logic level (e.g., 0.2V). When V_L exceeds V_{set}, the comparator output voltage goes to a low level which turns the transistor switch OFF and V_o is at a high logic level (e.g., +2.4V). These logic level voltages are fed to the digital engine controller to control fuel or both fuel and ignition. When V_o is low, fuel is ON; when V_o is high, fuel is OFF. If ignition is controlled also, it has the same conditions as fuel.

Actual wheel speed is fed
to an inverting amplifier
and lead term amplifier;
the resultant speed signal
is compared with V_{set} by a
voltage comparator. The
voltage comparator output
is used by the engine
controller to control fuel
metering and ignition.

Without fuel (or fuel and ignition), the engine drive force drops and the drive wheels slow down. The slow down is detected by the speed sensor and causes V_L to decrease. When V_L goes below V_{set}, the V_c flips high and restores the fuel (or fuel and ignition). This all happens very fast so the engine is still turning. When fuel (or fuel and ignition) is restored, the engine "restarts" and power is again delivered to the drive wheels.

**Figure 8-14.
Tire Slip Detector and
Fuel Flow Interruptor**

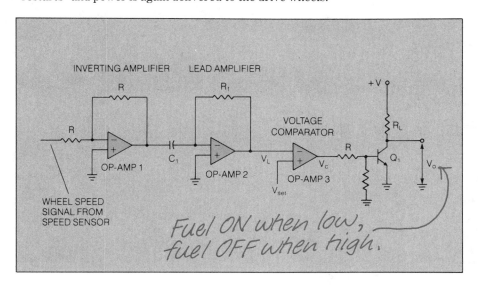

All of this action takes place so quickly that the drive wheels have little chance to slip and maximum drive force is obtained to move the vehicle. Tests have been run with cars equipped with tire slip controls on icy pavement. Colored bands were painted on the drive wheel tires and motion picture cameras recorded the tire motion during a hard acceleration (which normally would cause severe tire slip). The slow motion pictures showed the tires repeatedly breaking away, slowing down, re-grabbing the road surface and rolling a while, then breaking away again. This action repeated itself continuously as the vehicle accelerated.

Tire slip control systems were introduced originally where the control output transistor of *Figure 8-14* controlled the ignition and not the fuel. This system approach was discarded because the unused fuel would pass through the engine to the exhaust and cause excessive exhaust emissions.

BRAKE SKID CONTROL

A very similar set of electronics can be used in a system to prevent brake skid as shown in *Figure 8-15*. The difference is that in the brake skid system, the wheel motion stops (called wheel lock-up) and the vehicle skids. Therefore, deceleration of wheel speed is detected rather than acceleration. For this reason, the input signal to the inverting amplifier is applied to the + terminal rather than the − terminal in *Figure 8-14*. Wheel lock-up can be very dangerous, especially if the front wheels lock, because it reduces the driver's ability to steer the vehicle.

An electronic circuit similar to that used in a tire-slip controller also can be used in systems that prevent brake skid. Excessive wheel deceleration is used as a control signal to activate a valve that pulses the brake system.

**Figure 8-15.
Brake Skid Detector with
Brake Bypass Driver**

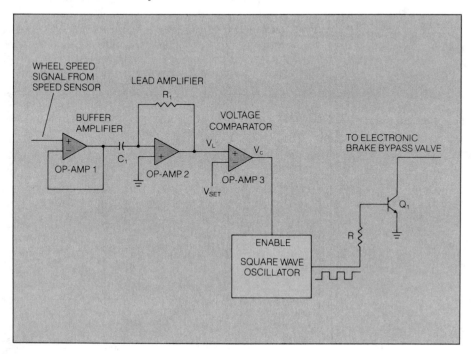

Another difference is in the output circuit. It is now a square-wave generator that has a 30 hertz (30 cycles per second) frequency. The oscillator is turned ON or OFF by the V_c output signal from the comparator. When there is a large deceleration, V_L is at a high voltage level, above V_{set}. This produces a high V_c which turns the square wave generator ON. The square wave signal is used to control the brakes intermittently, releasing momentarily the pressure that is applied, then causing it to be applied again. All this occurs at the 30 Hz rate. Under normal driving conditions, V_L is lower than V_{SET} which causes V_c to be low and holds the square wave generator OFF. Normal brake pressure is applied as needed when the square wave generator is OFF.

Test Results

Tests run on this system showed basically the same results as the tire slip tests. Slow motion films made under skid conditions showed the wheels repeatedly locked up and released until the car rolled to a safe stop.

In an initial system, a brake bypass valve was used to interrupt the brake line pressure. This system was discarded because of reliability and potential vehicle manufacturer's liability for accidents. Any system that controls a basic safety related system such as the brakes has to be extremely reliable and have a fail-safe mechanism. Although the brake skid controller did improve vehicle safety in certain driving conditions, the possibility of a system failure causing loss of brakes was determined to be more dangerous than loss of control due to skidding. Hopefully, new techniques will allow passenger safety systems of this type to be applied in the future.

SUMMARY

In this chapter, several electronic automotive motion controls have been discussed: a vehicle speed controller for cruise control, an engine governor, a tire slip controller and a brake skid controller. In each case, the electronic systems were shown and the important design and operating parameters were discussed. Next, we'll talk about instrumentation.

Quiz for Chapter 8

1. What does a system's time constant determine?
   a. The system low frequency gain
   b. The time it takes for the system to reach 63% of its final value
   c. The system's final value
   d. None of the above

2. What does a low pass filter do?
   a. Amplifies low frequencies and passes high
   b. Blocks low frequencies and passes high
   c. Blocks high frequencies and passes low
   d. a and c above

3. What kind of control electronics does the cruise control use?
   a. Proportional–Integral control
   b. Velocity feedback control
   c. Lead term control
   d. None of the above

4. How is the engine speed governor control electronics different from the cruise control electronics?
   a. The cruise control has a lead term
   b. The governor has a lead term
   c. Neither has a lead term
   d. a and b above

5. What other system uses a lead term?
   a. Tire slip control
   b. Cannister purge
   c. Cruise control
   d. None of the above

6. What type of damping usually produces the very best response?
   a. Under
   b. Optimally
   c. Over
   d. Critically

7. What value of damping coefficient produces a critically damped system?
   a. 12
   b. 3
   c. 1
   d. 0

8. Does a system's damping coefficient have anything to do with how wet the system is?
   a. Yes
   b. No
   c. Maybe
   d. Probably not

9. An optimally damped system has a damping coefficient of:
   a. Equal to 1
   b. Greater than 1
   c. Less than 1
   d. Greater than 10

10. If a system's open-loop gain is greater than one and the phase shift of the output from the input lags by 180°, the system is:
   a. Stable
   b. Does not oscillate
   c. Unstable and oscillates
   d. Open loop

Automotive Instrumentation

ABOUT THIS CHAPTER

Automotive instrumentation includes the equipment and devices which measure engine and other vehicle variables and display their status to the driver. From about the late 1920's until the late 1950's, the standard automotive instrumentation included the: speedometer, oil pressure gauge, coolant temperature gauge, battery charging-rate gauge, and fuel quantity gauge. Strictly speaking, only the latter two are electrical instruments. In fact, this electrical instrumentation was generally regarded as a minor part of the automotive electrical system. However, by the late 1950's, the gauges for oil pressure, coolant temperature, and battery charging rate were replaced by warning lights that were turned on only if specified limits were exceeded. This was done primarily to reduce vehicle cost and because of the presumption that many people did not read the gauges.

Low cost solid-state electronics, including microprocessors, display devices, and some sensors, have brought about major changes in automotive instrumentation.

This automotive instrumentation was not electronic until the 1970's. At that time, the availability of relatively low-cost solid-state electronics brought about a major change in automotive instrumentation. In this chapter, some of the electronic instrumentation presently available is described. The trend toward the use of electronics in instrumentation is continuing and the utilization of low-cost electronics is increasing with each new model year.

MODERN AUTOMOTIVE INSTRUMENTATION

The evolution of instrumentation in automobiles has been influenced by electronic technological advances in much the same way as the engine control system which we have discussed. In particular, the microprocessor, solid-state display devices, and some solid-state sensors have come into use. In order to put these developments into perspective, recall the general block diagram for instrumentation which is repeated in *Figure 9-1*.

In electronic instrumentation, a sensor is required to convert any nonelectrical signal to an equivalent voltage or current. Then, electronic signal processing is performed on the sensor output to produce an electrical signal which is capable of driving the display device. The display device is either an electromechanical device or an electro-optical device which can be read by the vehicle driver. If a quantity is to be measured that is already in electrical form, e.g. the battery charging current, then this signal can be used directly and no sensor is required.

**Figure 9-1.
General Instrumentation
Block Diagram**

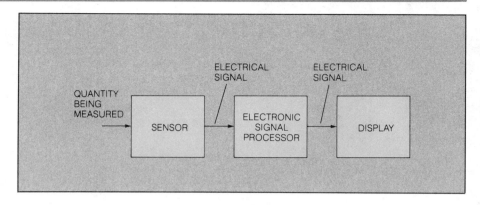

In some modern automotive instrumentation, a microcomputer performs all of the signal processing operations for several measurements. A block diagram for such an instrumentation system is shown in *Figure 9-2.*

**Figure 9-2.
Microcomputer Based
Signal Processing
System**

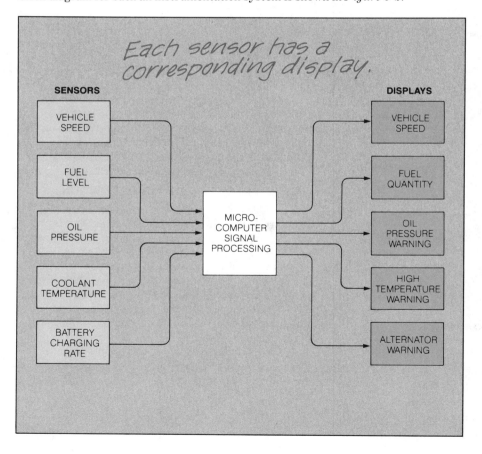

INPUT AND OUTPUT SIGNAL CONVERSION

Most sensors provide an analog output, while computers require digital inputs. A/D converters convert analog signals to digital codes appropriate for signal processing by the computer.

Generally speaking, the sensor output and computer input are not directly compatible. As we've discussed before, most automotive sensors have an analog output, but the computer requires a digital input. Most microcomputers have an 8-bit input.

The sensor analog output voltage can be made compatible with the computer/microprocessor by means of an analog-to-digital (A/D) converter as illustrated in *Figure 9-3*. The A/D converter generates the 8-bit binary number that represents the sensor output voltage. (Recall the discussion on A/D converters in Chapter 4.)

**Figure 9-3.
Analog-to-Digital
Conversion**

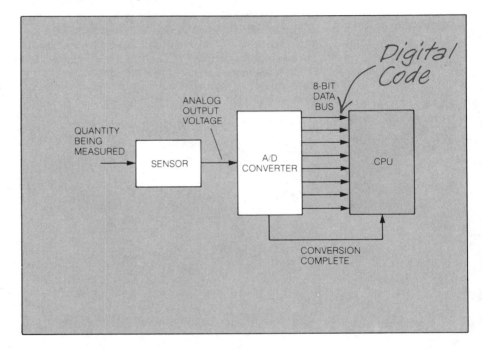

The conversion process requires some amount of time which depends primarily upon the A/D converter. When the binary number is ready, the A/D converter signals the computer that the conversion is completed and the data is ready to be entered into the computer for appropriate signal processing. The A/D converter signals the computer by the change of logic state on a separate lead (labeled "conversion complete" in *Figure 9-3*) connected between the computer and the A/D converter. (Also, recall the use of interrupts for this purpose as discussed in Chapter 4.)

The output voltage of each analog sensor for which the computer performs signal processing must be converted in this way. Once the conversion is complete, the digital output is transferred into a register in the computer.

The results of the signal processing (i.e. the computer output) is in a digital format. In most automotive applications, it is an 8-bit binary number. If the output is to drive a digital display, then this output can be used directly. However, if an analog display is used, the binary number must be converted to the appropriate analog signal with a digital to analog (D/A) converter.

Figure 9-4 illustrates the D/A converter in the computer output. The eight digital output leads transfer the results of the signal processing to a D/A converter. When the transfer is complete, the computer signals the D/A converter to start converting. The D/A output generates a voltage which is proportional to the binary number in the computer output. A capacitor is often connected across the D/A output to store the analog output between samples. The sampling of the sensor output, A/D conversion, digital signal processing, and D/A conversion all take place during the time slot alloted for the measurement of the variable in a sampling time sequence to be discussed shortly.

When an analog output signal is required to drive an analog display, a D/A converter is used. The DAC generates a voltage that is proportional to the binary number that the computer sends to the converter.

**Figure 9-4.
Digital-to-Analog
Conversion**

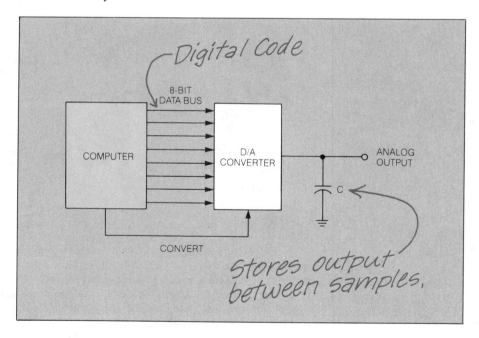

Switching and Reconfiguration

Of course, the computer can only deal with the measurement of a single quantity at any one time. Therefore, the computer input must be connected to only one sensor at a time and the computer output must be connected only to the corresponding display. The computer performs any necessary signal processing on that sensor signal and then generates an output signal to the appropriate display device.

We can understand this switching process by considering a functionally equivalent block diagram such as is shown in *Figure 9-5*. In this scheme, the various sensor outputs and display inputs are connected to a pair of multi-position rotary switches; one for the input and one for the output of the computer. The switches are mechanically connected such that they rotate together. Whenever the input switch connects the computer input to the appropriate sensor for measuring some quantity, the output switch connects the computer output to the corresponding display or warning device. Thus, with the switches in a specific position, the automotive instrumentation system corresponds to the block diagram shown in *Figure 9-1*.

The computer monitors each sensor individually, and provides output signals to its display component before going on to another sensor.

Figure 9-5.
Input/Output Switching
Scheme for Sampling

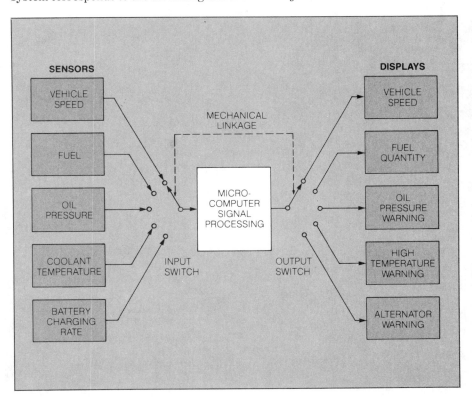

The switching of sensor and display inputs is performed with solid-state switches known as multiplexers and output switching by demultiplexers.

The computer can control the input and output switching operation. However, instead of a mechanical switch as shown in *Figure 9-5*, the actual switching is done by means of a solid-state electronic switching device called a multiplexer (MUX), shown in *Figure 9-6*, which selects one of several inputs for the output. The MUX has data select inputs that the computer activates to select the desired input signal. In other words, the computer "moves the switch" to connect itself to the desired sensor. Similarly, the output switching (which is often called demultiplexing or DEMUX) is performed with a MUX connected in reverse as shown in *Figure 9-7*. The MUX and DEMUX selection is controlled by the computer. Note that in *Figure 9-6* and *9-7* each bit of the digital code is multiplexed and demultiplexed.

Figure 9-6.
Data Multiplexer

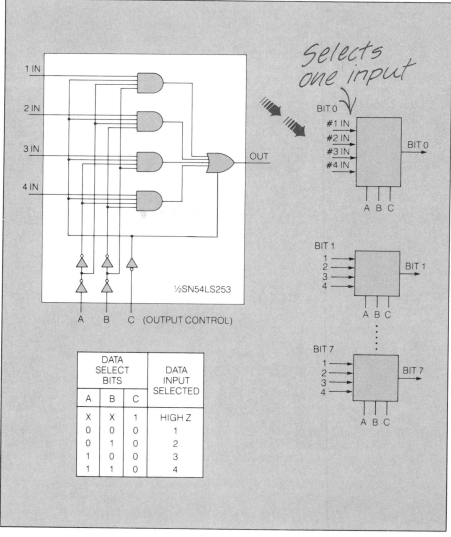

DATA SELECT BITS			DATA INPUT SELECTED
A	B	C	
X	X	1	HIGH Z
0	0	0	1
0	1	0	2
1	0	0	3
1	1	0	4

Figure 9-7.
Data Demultiplexer

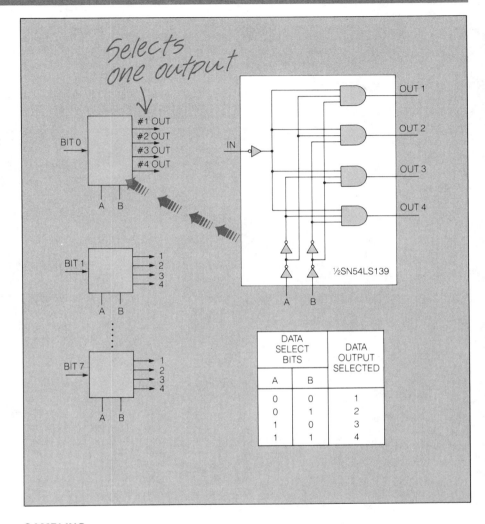

SAMPLING

Only one variable can be sampled by the computer at a time. The other variables must wait a set period of time before being sampled again by the computer.

The measurement of any quantity takes place only when the input and output switches (MUX and DEMUX) actually connect the corresponding sensor and display to the computer. However, there are several variables to be measured and displayed, but only one variable can be accommodated at any instant. Once a quantity has been measured, it must wait until the other variables have been measured before it is measured again.

This process of measuring a quantity intermittently is called sampling and the time between successive samples of the same quantity is called the sample period.

One possible scheme for measuring several variables by this process is to sample each quantity sequentially giving each measurement a fixed time-slot, t, out of the total sample period, T, as illustrated in *Figure 9-8*. This method is satisfactory as long as the sample period is small compared to the time in which any quantity changes appreciably. Certain quantities such as coolant temperature and fuel quantity change very slowly with time. For such variables, a sample period of a few seconds or longer is often adequate.

**Figure 9-8.
Sequential Sampling**

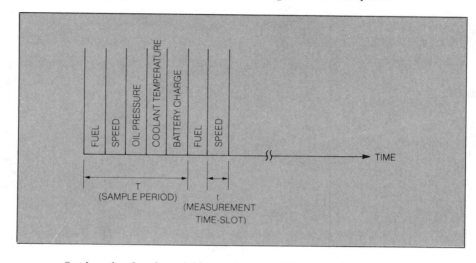

Some variables, such as speed and battery charge, change much faster than others. To effectively monitor these differences the computer uses different sampling times.

On the other hand, variables such as speed, battery charge, and fuel consumption rate change relatively quickly and require a much shorter sample period, perhaps every second or even tenths of a second. To accommodate the various rates of change of the automotive variables being measured, the sample period varies from one quantity to another. The most rapidly changing quantities are sampled with a very short sample period whereas those which change slowly are sampled with a long sample period.

In addition to sample period, the time slot allotted for each quantity must be long enough to complete the measurement and any A/D or D/A conversion required. The computer program is designed with all these factors in mind so that adequate time slots and sample periods are allowed for each variable. The computer then simply follows the program schedule.

System Configurations

Because the various types of sensors provide different output levels, the microcomputer must be reconfigured for the characteristics of each sensor. This reconfiguration is accomplished through the instructions contained in the computer program.

We have seen that the various sensors produce outputs that differ from one another and that different signal processing is required to produce the required outputs. Therefore, the signal processor must be changed or reconfigured for each different switch position. *Figure 9-9* shows the system configuration for various types of display. The only practical means of achieving this reconfiguration of the signal processing operation is by means of a microcomputer; it is possible to perform a wide range of signal processing operations by an appropriate combination of arithmetic or logical operations that result from the program stored in the computer's memory.

**Figure 9-9.
Configuration for
Various Types of Display**

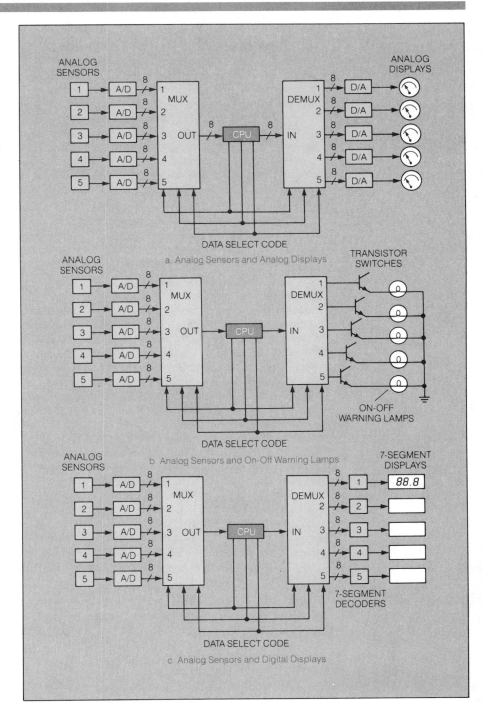

ANALOG
SENSORS

ANALOG
DISPLAYS

a. Analog Sensors and Analog Displays

TRANSISTOR
SWITCHES

ON-OFF
WARNING LAMPS

b. Analog Sensors and On-Off Warning Lamps

7-SEGMENT
DISPLAYS

7-SEGMENT
DECODERS

c. Analog Sensors and Digital Displays

Advantages of Computer-Based Instrumentation

Because the configuration is under control of a computer program, computer-based instrumentation has great flexibility. To change from the instrumentation for one vehicle or one model to another requires only a change of computer program. This change can often be implemented by replacing one ROM with another. (Remember that the program is permanently stored in a read-only memory that is typically packaged in a single plug-in IC package.)

Yet another benefit of the microcomputer based electronic automotive instrumentation is improved performance compared to the conventional instrumentation. In the electronic instrumentation, measurement errors are much smaller than for the conventional instrumentation. For example, the conventional fuel gauge system has errors which are associated with variations in the:

1. mechanical characteristics of the tank,
2. sender unit,
3. instrument voltage regulator, and
4. indicator (galvanometer).

Computer-based instrumentation is more accurate than conventional instrumentation. And due to the computer's program they are more easily changed.

The electronic system completely eliminates the error which results from imperfect voltage regulation. Generally speaking, the electronic fuel quantity measurement maintains calibration over essentially the entire range of automotive electrical system conditions. Moreover, it significantly improves the indicator accuracy by replacing the electromechanical galvanometer indicator with an all-electronic digital display. While the impact of these improvements vary with the vehicle and fuel level, a typical electronic system has about 44% less error at half-a-tank than a conventional system.

There is a similar improvement in the electronic speedometer compared to the conventional electromechanical speedometer. The electronic system has about 35% less variation than the conventional system.

There is yet another benefit gained by using microcomputer based instrumentation. Because of its flexibility and computational capability, the computer can manipulate and combine various sensor inputs to provide outputs other than the output originally desired from a measurement. This capability greatly enhances the set of useful quantities which can be displayed to the driver. For example, a typical computer based instrumentation system can compute the amount of miles that can be driven on the remaining fuel for the current set of steady operating conditions. We will discuss the details of this computation in a later section of this chapter. The point to be made here is that this output information is a "by-product" of inputs whose primary purpose is for other outputs.

FUEL QUANTITY MEASUREMENT

During a measurement of fuel quantity, the MUX switch connects the computer input to the fuel quantity sensor as shown in *Figure 9-10*. This sensor output is converted and then sent to the computer for signal processing.

Some fuel quantity sensors use a float within the fuel tank; the float is mechanically linked to a potentiometer which operates as a voltage divider.

Several fuel quantity sensor configurations are available. *Figure 9-11* illustrates the type of sensor we will describe.

Normally, the sensor is mounted so that the float remains laterally near the center of the tank for all fuel levels. A constant current passes through the sensor potentiometer since it is connected directly across the regulated voltage source. The potentiometer is used as a voltage divider so that the voltage at the wiper arm is related to the float position, which is determined by fuel level.

**Figure 9-10.
Fuel Quantity
Measurement**

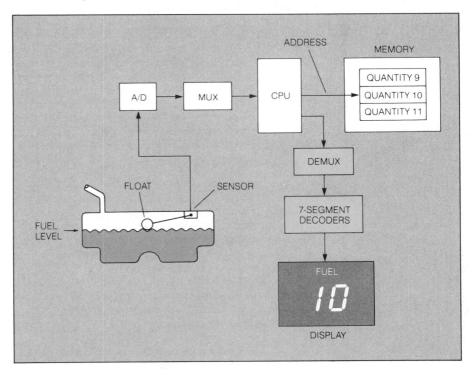

**Figure 9-11.
Fuel Quantity Sensor**

Voltage level represents fuel level.

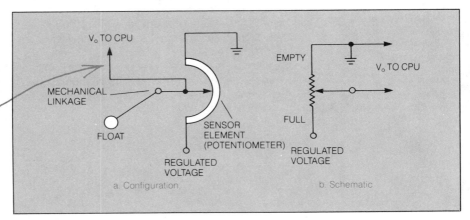

The fuel sensor output voltage is converted into a binary code by an ADC. The computer compares this code to codes stored in a look up table that correspond to actual fuel quantities for the specific fuel tank.

The computer compensates for fuel slosh by averaging float sensor readings over a period of time.

Because of the complex shape of the fuel tank, the sensor output voltage is not directly proportional to fuel quantity in gallons. The computer memory contains the relationship between sensor voltage (in binary number equivalent) and fuel quantity for the particular fuel tank used on the vehicle.

The computer reads the binary number from the A/D converter that corresponds to sensor voltage and uses it to address a particular memory location. Another binary number which corresponds to the actual fuel quantity in gallons for that sensor voltage is stored in that memory location. The computer then uses the number from memory to generate the appropriate display voltage; i.e., either analog or digital depending upon display type, and sends that signal via DEMUX to the display.

The computer based signal processing can also compensate for fuel slosh. As the car moves over the road, the fuel sloshes about and the float bobs up and down around the average position which corresponds to the correct level for a stationary vehicle. The computer compensates for slosh by computing a running average. It does this by storing several samples over a few seconds and computing the arithmetic average of the sensor output. The oldest samples are continually discarded as new samples are obtained. The averaged output is used as the memory address as described above.

COOLANT TEMPERATURE MEASUREMENT

Another important automotive parameter which is measured by the instrumentation is the coolant temperature. The measurement of this quantity is different from fuel quantity measurements because usually it is not important for the driver to know the actual temperature at all times. Rather, for safe operation of the engine, the driver only needs to know that the coolant temperature is less than a critical value. A block diagram of the measuring system is shown in *Figure 9-12*.

The coolant temperature sensor used in most cars is a solid state sensor called a thermistor. Recall that we discussed this type of sensor in Chapter 5. The resistance of this sensor decreases with temperature. *Figure 9-13* shows the circuit connection and a sketch of a typical sensor output voltage versus temperature curve.

To measure coolant temperature, the sensor output from a thermistor is converted into a digital signal and compared to a maximum safe value stored in memory. If the maximum value is exceeded, the computer generates a signal to trigger a temperature warning lamp.

The sensor output voltage is sampled during the appropriate time slot and is converted to a binary number equivalent by the A/D converter. The computer compares this binary number to the one stored in memory that corresponds to the high temperature limit. If the coolant temperature exceeds the limit, then an output signal is generated which activates the warning indicator. If the limit is not exceeded, then the output signal is not generated and the warning message is not activated. A proportional display of actual temperature can be used if the memory contains a cross reference table between sensor output voltage and the corresponding temperature, similar to that described for the fuel quantity table.

**Figure 9-12.
Coolant Temperature
Measurement**

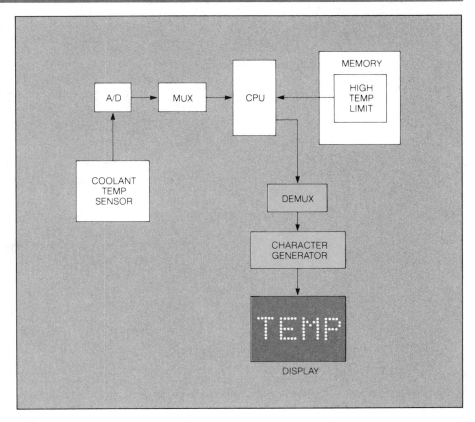

**Figure 9-13.
Coolant Temperature
Sensor**

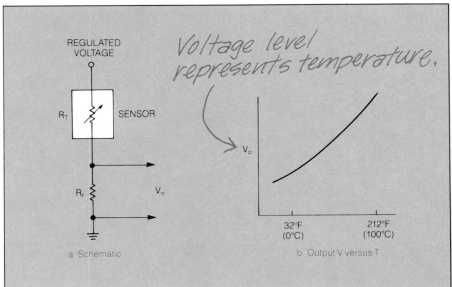

OIL PRESSURE MEASUREMENT

Another important function of the automotive instrumentation is the measurement of engine oil pressure. Whenever the oil pressure goes outside allowable limits, a warning message is displayed to the driver. This function is similar in many respects to the high coolant temperature warning function.

In the case of oil pressure, it is important for the driver to know whenever oil pressure falls below a lower limit. It is also possible for oil pressure to go above an allowable upper limit; however, many manufacturers do not include high oil pressure warning in the instrumentation.

An oil pressure warning system is illustrated in *Figure 9-14*. A variable resistance oil pressure sensor as shown in *Figure 9-15* is used and a voltage is developed across a series fixed resistance which is proportional to oil pressure.

Oil pressure warning systems use a variable resistance sensor as part of a voltage divider. This arrangement provides a varying voltage that corresponds to changes in oil pressure.

**Figure 9-14.
Oil Pressure
Measurement**

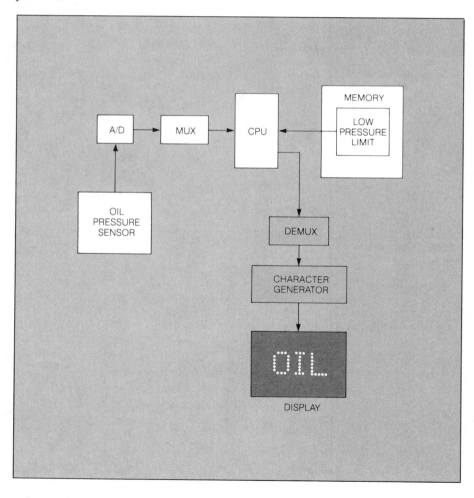

Figure 9-15.
Oil Pressure Sensor

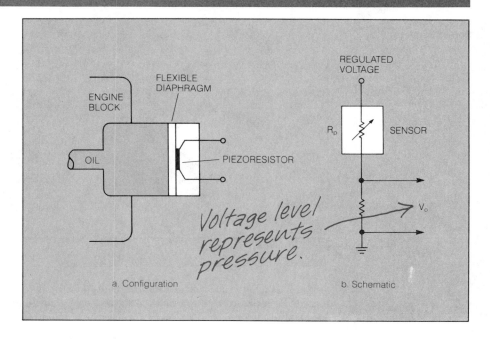

a. Configuration   b. Schematic

To sample oil pressure, the sensor output is converted to a binary code and compared with codes representing the safe limits for oil pressure. If the sampled code falls outside these limits, the computer triggers the oil warning display.

During the time-slot for measuring, the oil pressure sensor voltage is sampled through the MUX switch and converted to a binary number in the A/D converter. The computer reads this binary number and compares it with the binary number in memory for the allowed oil pressure limits. If the oil pressure is below the allowed lower limit or above the allowed upper limit, then an output signal is generated which activates the oil pressure warning light through the DEMUX.

It is also possible to use a proportional display of actual oil pressure if a cross-reference table similar to the fuel quantity table is used. A digital display can be driven directly from the computer. An analog display, such as the electric gauge, requires a D/A converter.

VEHICLE SPEED MEASUREMENT

In one type of vehicle speed measurement system, the vehicle speed information is mechanically coupled to the speed sensor by a flexible cable from the driveshaft which rotates at an angular speed proportional to vehicle speed. A speed sensor driven by this cable generates a pulsed electrical signal (*Figure 9-16*) which is processed by the computer to produce an output to drive a display to indicate speed as shown in *Figure 9-17*.

Speed sensors are mechanically linked to the vehicle's drive train. Such sensors produce a pulsed output signal by causing a slotted disk to rotate between a light source and a light detector.

The flexible cable drives a slotted disk which rotates between a light source and a light detector. The light detector produces an output voltage when exposed to light from the light source on the other side of the disk. The placement of the source, disk, and detector is such that the slotted disk interrupts or passes the light from source to detector depending upon whether a slot is in the line of sight from source to detector.

**Figure 9-16.
Speed Sensor**

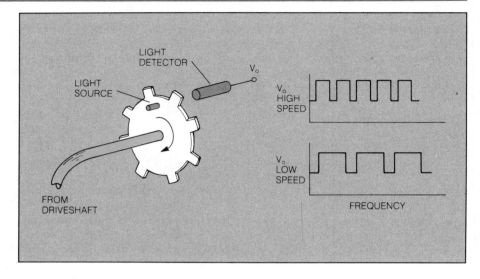

The light detector generates an output pulse of voltage whenever a pulse of light passes through a slot to the detector. The number of pulses which are generated per second is proportional to the number of slots in the disk and the vehicle speed.

$$f = N\,S\,K$$

where

f = frequency in pulses per second,
N = number of slots in the sensor disk,
S = vehicle speed, and
K = proportionality constant which accounts
for differential gear ratio and wheel size

The output pulses are passed through a sample gate to a digital counter as shown in *Figure 9-17*. The sample gate is an electronic switch which either passes the pulses to the counter or does not pass them. The time interval that the gate is open is precisely controlled by the computer. The digital counter counts the number of pulses from the light detector during the time, t, the gate is open. The number of pulses which are counted by the digital counter is given by:

$$P = t\,N\,S\,K$$

That is, the number P, is proportional to vehicle speed S.

The computer reads the number P, then resets the counter to zero to prepare it for the next count. After performing computations and filtering, the computer generates a signal for the display to indicate the vehicle speed. A digital display can be directly driven by the computer. Either MPH or KM/H may be selected. If an analog display is used, a D/A converter must drive the display. Both MPH and KM/H usually are calibrated on the analog scale.

Pulses from the speed sensor are counted during a specific sampling time by a digital counter. The computer translates the number of pulses into a corresponding speed.

Figure 9-17.
Vehicle Speed
Measurement

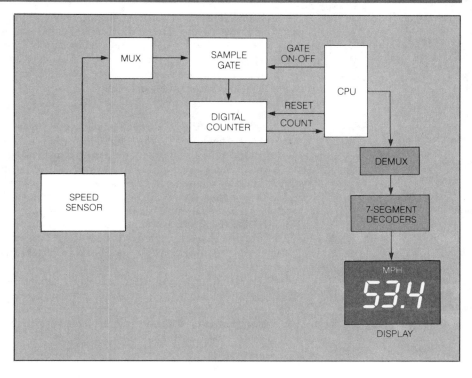

DISPLAY DEVICES

One of the most important components of any measuring instrument is the display device. In automotive instrumentation, the display device must present the results of measurement to the driver in a form which is easy to read and understand. The first automotive electronic displays were electromechanical devices. The speedometer, ammeter, and fuel quantity gauge were this type. Then many manufacturers began using warning lamps instead of gauges to cut cost. A warning lamp can be considered as a type of electro-optical display.

Electromechanical and simple electro-optical displays are being replaced by sophisticated electronic displays that provide the driver with numeric or alphabetic information.

Recent developments in solid-state technology in the field called optoelectronics have lead to much more sophisticated electro-optical display devices which are capable of indicating alphanumeric data[1]. This permits both numeric and alphabetic information to be used to display the results of measurements of automotive variables or parameters. This capability allows messages in English or other languages to be given to the driver. For these devices, the input is an electronic digital signal. This makes them compatible with computer-based instrumentation whereas electromechanical displays require a D/A converter.

[1]L.B. and B.R. Masten, *Understanding Optronics*, Texas Instruments Incorporated, Dallas, Texas, 1981

Automobile manufacturers considered several types of electronic displays for automotive instrumentation, but only these three were ever seriously considered: light-emitting diode (LED), liquid crystal display (LCD), and vacuum fluorescent (VF). It now appears that the latter will be the predominant type for at least the near future. We will discuss each type briefly to explain the use in automotive applications.

LED

LED's are semiconductors that emit light when current is passed through the diode. LED's are difficult to view in bright sunlight.

The light-emitting diode is a semiconductor diode that is constructed in a manner and of a material so that light is emitted when an electrical current is passed through it. The semiconductor material most often used for an LED that emits red light is gallium arsenide phosphide. Light is emitted at the diode's PN junction when the positive carriers combine with the negative carriers at the junction. The diode is constructed so that the light generated at the junction can escape from the diode so it can be seen.

An LED display is normally made of small dots or rectangular segments arranged so that numbers and letters can be formed when selected dots or segments are turned on. The LED display was the first display used in electronic digital watches.

The LED is not well suited for automotive display use because of its low brightness. Although it can be seen easily in darkness, it is difficult-to-impossible to see in bright sunlight. It also requires more electrical power than an LCD display; however, its power requirements are not great enough to be a problem for automotive use.

LCD

LCD's use a liquid that possesses the ability to rotate the polarization of polarized light. LCD displays have low power requirements.

The heart of an LCD is a special liquid which is called a twisted nematic liquid crystal. This liquid has the capability of rotating the polarization of linearly polarized light. The LCD display is commonly used in electronic digital watch displays because of its extremely low electrical power and relatively low voltage requirements.

Linearly polarized light has all of the vibration of the optical waves in the same direction. Light from the sun and most artificial light sources is not polarized and the waves vibrate randomly in many directions.

Non-polarized light can be polarized by passing the light through a polarizing material. To illustrate; think of a picket fence with narrow gaps between pickets. If you pass a rope between two pickets and whip one end of the rope up and down, the ripples in the rope will pass through the fence. The ripples represent light waves and the picket fence represents a polarizing material. If you whip the rope in any other direction other than vertically, the ripples will not pass through.

Now visualize another picket fence turned 90° so the pickets are horizontal. Place this fence behind the vertical picket fence. This arrangement is called a cross-polarizer. If you whip the rope now in any direction, no ripples will pass through both fences. Similarly, if a cross-polarized lens is used for light, no light will pass through the lens.

The configuration of an LCD can be understood from the schematic drawing of *Figure 9-18*. The liquid crystal is sandwiched between a pair of glass plates which have transparent electrically conductive coatings. The transparent conductor is deposited on the front glass plate in the form of the character or segment of a character which is to be displayed. Next a layer of dielectric (insulating) material is coated on the glass plate which produces the desired alignment of the liquid crystal molecules. The polarization of the molecules is vertical at the front and they gradually rotate through the liquid crystal structure until the molecules at the back are horizontally polarized. Thus, the molecules of the liquid crystal rotate 90 degrees from the front plate to the back plate so their polarization matches that of the front and back polarizers with no voltage applied.

**Figure 9-18.
Typical LCD
Construction**

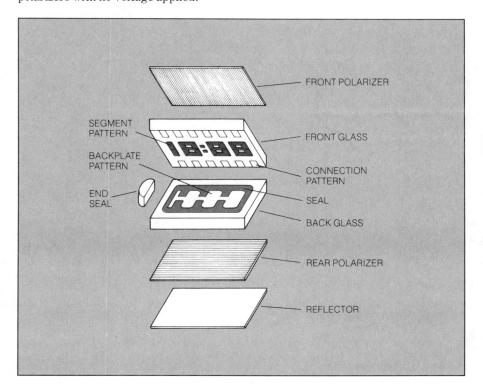

When current is not being applied to an LCD display, light entering the crystal is polarized by the front polarizer, rotated, passed through the rear polarizer and then reflected off the reflector. The reflected light causes the segment to appear blank.

The operation of the LCD in the absence of applied voltage can be understood with reference to *Figure 9-19a*. Ambient light enters through the front polarizer so the light entering the front plate is vertically polarized. As it passes through the liquid crystal, the light polarization is changed by the orientation of the molecules. When the light reaches the back of the crystal, its polarization has been rotated 90 degrees so it is horizontally polarized. The light is reflected from the reflector at the rear. It passes back through the liquid crystal structure, the polarization again being rotated, and passes out of the front polarizer. Thus, a viewer sees reflected ambient light.

**Figure 9-19.
Liquid Crystal
Polarization**

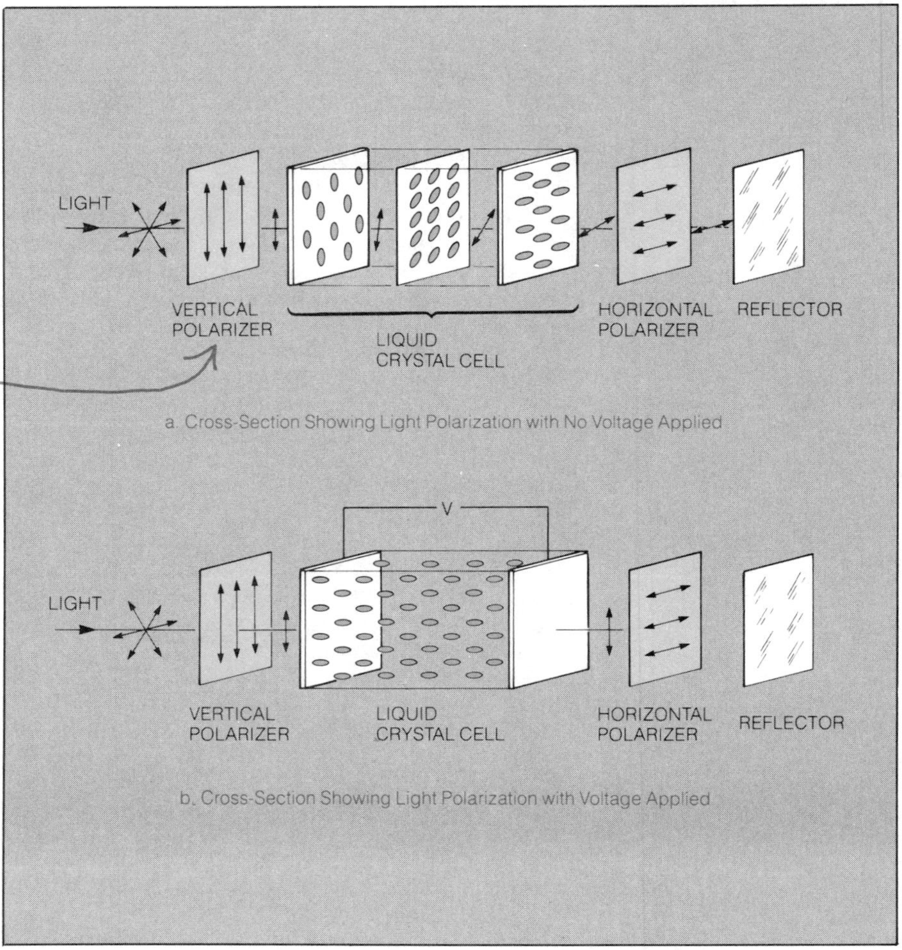

Only pass light in a specific direction.

a. Cross-Section Showing Light Polarization with No Voltage Applied

LIGHT

VERTICAL
POLARIZER

LIQUID
CRYSTAL CELL

HORIZONTAL
POLARIZER

REFLECTOR

b. Cross-Section Showing Light Polarization with Voltage Applied

When a voltage is applied to a display segment, the crystal's molecules change and do not rotate the polarized light. Since the light cannot align with the rear polarizer, it is not reflected, and the segment appears dark.

The effect of an applied voltage to the transmission of light through this device can be understood from *Figure 9-19b*. A voltage applied to any of the segments of the display causes the liquid crystal molecules under that segment only to be aligned in a straight line rather than twisted. In this case, the light which enters the liquid crystal in the vicinity of the segment passes through the crystal structure without the polarization being rotated. Since the light has been vertically polarized by the front vertical polarizing plate, the light is blocked by the horizontal polarizer so it cannot reach the reflector. Thus light which enters the cell in the vicinity of the energized segment is not returned to the front face. This segment will appear dark to the viewer, the surrounding area will be light, and the segment will be visible in the presence of ambient light.

The LCD is an excellent display device because of its low power requirement and relatively low cost. However, a big disadvantage of the LCD for automotive application is the need for an external light source for viewing in the dark. Its characteristic is just opposite that of the LED; that is, the LCD is readable in the daytime, but not at night. For night driving, the display must be illuminated by small lamps inside the display. Another disadvantage is that the display does not work well at low temperatures that may be encountered during winter driving in some areas. These characteristics of the LCD have limited its use in automotive instrumentation.

VFD

VFD's use a phosphor material that emits light when bombarded by electrons. VFDs provide readability over a wide range of conditions.

One of the most common automotive display devices in use today is the vacuum fluorescent display (VFD). This device generates light in much the same way as a television picture tube. That is, a phosphor material emits light when it is bombarded by energetic electrons. The display uses a heated filament coated with material that generates 'free' electrons when hot. The electrons are accelerated toward the anode by a relatively high voltage. When these high speed electrons strike the phosphor material on the anode, the phosphor material emits light. Most VF displays have a phosphor material that emits a blue-green light which provides good readability in the wide range of ambient light conditions that are present in an automobile.

The numeric characters are formed by shaping the anode segments in the form of a standard 7-segment character. The basic structure of a typical VFD is depicted in *Figure 9-20*. The filament is a special type of resistance wire and is heated by passing an electrical current through it. The coating on the heated filament produces free electrons which are accelerated by the electric field produced by a voltage on the accelerating grid. This grid consists of a fine wire mesh that allows the electrons to pass through. The electrons pass through because they are attracted to the anode which has a higher voltage than the grid. The high voltage is applied only to the anode of the segments needed to form the character to be displayed. The instrumentation computer selects the set of segments which are to emit light for any given message.

VFD brightness can be controlled by varying the voltage on the accelerator grid. As the accelerator grid voltage increases, the electrons strike the phosphor with greater intensity, resulting in increased light output.

Since the ambient light in an automobile varies between sunlight and darkness, it is desirable to adjust the brightness of the display in accordance with the ambient light. The brightness is controlled by varying the voltage on the accelerating grid. The higher the voltage, the harder the electrons strike the phosphor, and the brighter the light. *Figure 9-21* shows the brightness characteristics for a typical VFD device. A brightness of 200 fl (foot lamberts) might be selected on a bright sunny day whereas the brightness might only be 20 fl at night. The brightness can be set manually by the driver or automatically. In the latter case, a photoresistor is used to vary the grid voltage in accordance with the amount of ambient light. A photoresistor is a device whose resistance varies in proportion to the amount of light striking it.

**Figure 9-20.
Simplified Vacuum
Fluorescent Display
Configuration**

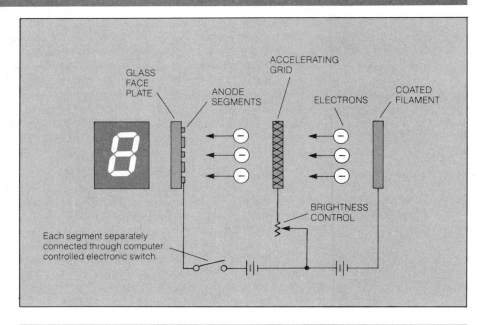

**Figure 9-21.
Brightness Control
Range for VF Display**

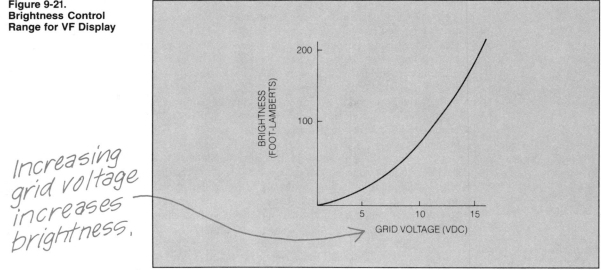

*Increasing
grid voltage
increases
brightness.*

The VFD display is likely to be the most widely used for some time. It operates with relatively low power and has stable operation over a wide temperature range. The most serious drawback for automotive application is its susceptibility to failure due to vibration and mechanical shock. However, this problem can be reduced by mounting the display on a shock absorbing isolation mount.

TRIP INFORMATION COMPUTER

One of the most popular electronic instruments for automobiles is the trip information system. This system has a number of interesting functions and can display several useful pieces of information to the driver including:

1. Present fuel economy
2. Average fuel economy
3. Average speed
4. Present vehicle location (relative to total trip distance)
5. Total elapsed trip time
6. Fuel remaining
7. Miles to empty fuel tank
8. Estimated time of arrival
9. Time of day
10. Engine RPM
11. Engine temperature
12. Average fuel cost per mile

There are other functions which can be performed and no doubt will be part of future developments. However, we will discuss a representative system having features which are common to most available systems.

A block diagram of this system is shown in *Figure 9-22*. Not shown in the block diagram is MUX, DEMUX and A/D converter components which are normally part of a computer-based instrument. This system can be implemented as a set of special functions of the main automotive instrumentation system or it can be a stand-alone system employing its own computer.

The vehicle inputs to this system come from the three sensors which measure the:

1. quantity of fuel remaining in the tank,
2. instantaneous fuel flow rate, and
3. vehicle speed.

Other inputs which are obtained by the computer from other parts of the control system are:

1. odometer mileage, and
2. time-of-day clock in the computer.

The driver enters inputs to the system through the keyboard. At the beginning of a trip, the driver initializes the system and enters the total trip distance and fuel cost. At any time during the trip, the driver can use the keyboard to ask for information to be displayed.

Trip information computers analyze fuel flow, vehicle speed, and fuel tank quantities, and then calculate information such as miles-to-empty, average fuel economy, and estimated arrival times.

Figure 9-22.
Trip Information System

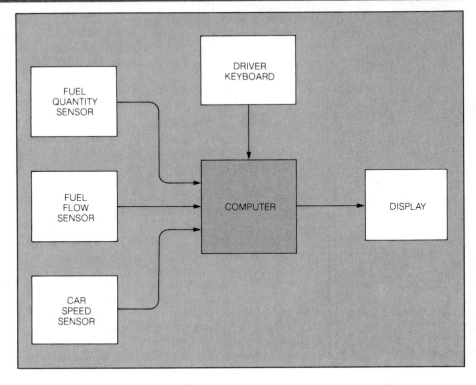

The system computes a particular trip parameter from the input data. For example, fuel ecomony in miles-per-gallon (MPG) can be found by computing:

$$MPG = S/F$$

where

S = speed in miles per hour, and
F = fuel consumption rate in gallons per hour.

As operating conditions change, the values provided by a trip information computer may also change.

Of course, this computation varies markedly as operating conditions vary. At a steady cruising speed along a level highway with a constant wind, fuel economy would be constant. If the driver were to then depress the accelerator (e.g. to pass traffic), the fuel consumption rate would temporarily increase faster than speed and MPG would be reduced for that time.

Another important trip parameter which this system can display is the distance to empty fuel tank D. This can be found by calculating:

$$D = MPG \times Q$$

where

Q = the quantity of fuel remaining in gallons.

Since D depends on MPG, it also changes as operating conditions change, but it is most useful for steady cruise along a highway.

Still another pair of parameters which can be calculated and displayed by this system are distance to destination, D and estimated time of arrival, ETA. These can be found by computing:

$$D = D_T - D_P$$
$$ETA = D/S$$

where

D_T = trip distance (entered by the driver)
D_P = present position (in miles traveled since start)
S = present vehicle speed

The computer can calculate the present position D_P by subtracting the start mileage D_I (from the odometer when the trip computer was initialized by the driver) from the present odometer mileage.

The average fuel cost per mile C can be found by calculating:

$$C = (D_p/MPG) \times FUEL\ COST\ PER\ GALLON$$

We have described only some of the more common functions of a trip information system. Many other useful and interesting operations are performed by the variety of systems which are available. Actually, the number of such functions which can be performed is limited primarily by cost and the availability of sensors to measure the required variables.

AUTOMOTIVE DIAGNOSTICS

The instrumentation computer can also be used as a diagnostic aid, during vehicle manufacturing, operation, or repair.

In certain automobile models, the instrumentation computer can perform the important function of diagnosis of the electronic engine control system. This diagnosis takes place at several different levels. One level is used during manufacturing to test the system. Another level is used by mechanics or interested car owners to diagnose engine control system problems. Some levels operate continuously and others are available only upon request from the keyboard.

In the continuous monitor mode, the diagnosis takes place under computer control. The computer activates connections to the vehicle sensors and looks for an open- or short-circuited sensor. If such a condition is detected, then a failure warning message is given to the driver on the alphanumeric display or by turning on a labeled warning light.

Another level of diagnosis which is available on some automobiles is capable of system diagnosis with much greater precision than the sensor open-circuit or short-circuit test. This level can identify specific system faults and can greatly speed up and simplify repair. This diagnostic level is normally used by mechanics, but can also be used by the owner if he is interested in repairing his own car.

Advanced diagnostics levels can perform a number of tests on a vehicle's electronics and sensors. These diagnostics are stored in the computer program, and can be accessed by a mechanic for repair purposes.

This diagnostic scheme is requested from the computer by the owner or mechanic by operating specific keyboard keys in a certain pattern. The diagnostic routine is actually a part of the computer program and can be conducted in a fully automatic mode or by manual sequencing (depending upon the system design).

During this diagnostic routine, a number of tests are performed on the system. One set of tests examines the output from each sensor. In certain models, the actual value of the sensor output is displayed. The mechanic can compare this value with the expected or acceptable values for this sensor. Even a partial failure of a sensor can be detected by this diagnostic procedure.

It is possible for the instrumentation to be designed such that the computer runs this performance test on all sensors automatically. The failure of any sensor to pass this diagnostic test will cause the computer to generate a warning message on the display. Of course, the instrumentation computer itself must be functioning to perform this test.

Quiz for Chapter 9

1. What is the primary purpose of automotive instrumentation?
   a. to indicate to the driver the value of certain critical variables and parameters
   b. to extend engine life
   c. to control engine operation
   d. entertainment of passengers

2. What are the 3 functional components of electronic instrumentation?
   a. sensor, MAP, display
   b. sensor, signal processing, error amplifier
   c. display, sensor, signal processing
   d. none of the above

3. What is the function of a multiplexer in computer based instrumentation?
   a. it measures several variables simultaneously
   b. it converts sensor analog signals to digital format
   c. it sequentially switches a set of sensor outputs to the instrumentation computer input

4. What is sampling?
   a. a signal processing algorithm
   b. a selective display method
   c. a method of measuring a continuously varying quantity at discrete time instants
   d. the rate of change of battery voltage

5. What is an A/D converter?
   a. a device which changes a continuously varying quantity to a digital format
   b. an 8-bit binary counter
   c. an analog-to-decimal converter
   d. a fluid coupling in the transmission

6. What type of sensor is commonly used for fuel quantity measurement?
   a. a thermistor
   b. a strain gauge
   c. a potentiometer whose movable arm is connected to a float
   d. a piezo-electric sensor

7. How is coolant temperature measured?
   a. with a mercury bulb thermometer
   b. with a strain gauge
   c. with a thermistor as a sensor
   d. with a galvanometer

8. What type of sensor is used to measure oil pressure?
   a. a pressure sensitive diaphragm
   b. an LVDT
   c. a wheatstone bridge
   d. a calorimeter

9. What is the predominant type of automotive digital display?
   a. light-emitting diode
   b. galvanometer
   c. vacuum fluorescent
   d. liquid crystal

10. What sensor input variables are used in a typical "Trip Computer" system?
    a. manifold pressure and engine speed
    b. RPM, barometric pressure and fuel quantity remaining
    c. MPG and fuel consumption
    d. car speed, fuel flow rate, fuel quantity remaining in tank

11. What does engine control diagnostics mean?
    a. a systematic procedure for finding faults in the electronic engine control system
    b. repair of the vehicle by a licensed mechanic
    c. measurement of engine output variables
    d. entry of data into the computer via a keyboard

12. Automatic electronic diagnosis is performed by:
    a. a special computer program
    b. all sensor inputs
    c. all actuator outputs
    d. none of the above

13. The term MUX refers to:
    a. an electronic switch which
        selects one of a set of inputs
        per an input code
    b. a digital output device
    c. a time slot
    d. none of the above

14. A D/A converter:
    a. is a disk access device
    b. converts the digital output of an
        instrumentation computer to
        analog form
    c. stores analog data
    d. enters digital data in a computer

15. In electronic instrumentation, fuel
quantity is displayed by:
    a. an ammeter
    b. a potentiometer
    c. a digital display
    d. none of the above

16. In an optical speed sensor, the
output frequency is given by:
    a. SK/n
    b. NS^2K
    c. NSK
    d. $tNSK$

17. The term LED refers to:
    a. level-equalizing detector
    b. light-emitting diode
    c. liquid crystal display
    d. none of the above

18. An LC display uses:
    a. a nematic liquid
    b. an incandescent lamp
    c. large electrical power
    d. a picket fence

19. Light is produced in a VFD by:
    a. ionic bombardment of a filament
    b. ambient temperature
    c. bombardment of a phosphor
        material by energetic electrons
    d. chemical action

20. Fuel economy is calculated in a trip
computer by:
    a. SF
    b. F/S
    c. S/F
    d. none of the above

Future Automotive Electronic Systems

ABOUT THIS CHAPTER

Up to this point in this book, we have been discussing automotive electronic technology of the recent past or present. The systems which we have discussed have been utilized at one time or another in one mode or another. Many of the electronic systems have been based upon technology of the late 1960's or early 1970's.

In this chapter, we will speculate about the future of automotive electronic systems. Some concepts are only laboratory studies and may not at the time of this writing have had any vehicle testing at all. Some of the system concepts have been or are being tested experimentally. Some are operating on a limited basis in automobiles.

Whether or not any of these concepts ever reaches a production phase will depend largely upon its technical feasibility and marketability. Some will simply be too costly to have sufficient customer appeal and will be abandoned by the major automobile manufacturers. However, some of these abandoned system concepts may be picked up by small electronics companies who manufacture and sell the product in the "after market" where the product is bought and installed by car owners or specialty shops.

On the other hand, one or more of these ideas may become a major market success and will be included in many models of automobiles. Some of these systems may even provide a significant selling point for one of the large automobile manufacturers.

We will present a summary of the major electronic systems which have been considered and which may be considered for future automotive application. For convenience, we will separate these ideas into the following categories:

1) Engine and Drivetrain
2) Safety
3) Instrumentation

ENGINE AND DRIVETRAIN

Several electronic systems have been considered and are under various stages of development which are applicable to the engine or drivetrain.

Ignition

One of the important automotive electronic systems having great potential for future application is the distributor-less ignition system. We discussed electronic ignition systems in Chapters 5 and 6, but recall that these systems still use the mechanical switching of the high voltage by the distributor. The distributor-less system uses an all-electronic system to replace the distributor as a means of switching ignition high voltage from one spark plug to another. Some of the motivations for eliminating the distributor are reduced initial cost, improved reliability and reduced service cost.

We can illustrate the concept of all-electronic ignition with the configuration shown in *Figure 10-1*. This circuit is applicable to a four-cylinder engine having, of course, four spark plugs to fire. This system actually fires a pair of spark plugs simultaneously (i.e. number 1 and number 4 fire together and number 2 and number 3 fire together). These cylinders are arranged such that whenever one cylinder of a pair is in the compression stroke and is ignited by the spark, the other cylinder of the pair is in its exhaust stroke. The spark has no effect upon this latter cylinder since there is no fuel mixture to ignite.

An all-electronic ignition system replaces the conventional distributor with an electronic switching circuit. The computer provides timed trigger pulses to a transistor network that fires spark plugs in pairs.

**Figure 10-1.
Distributor-less
Electronic Ignition**

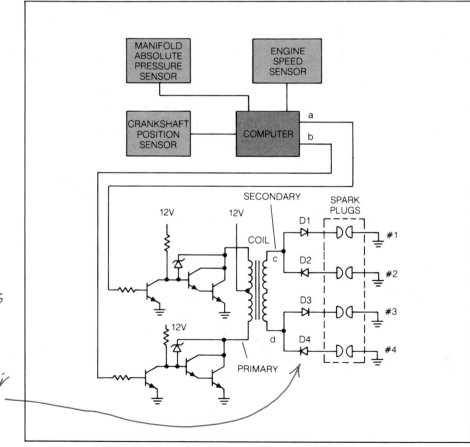

Reverse bias of diodes permit one spark plug of each pair to fire.

In this system, the position of the pistons is detected by the crankshaft position sensor as we've discussed for other systems. The computer determines the optimum ignition timing based on measurements of manifold absolute pressure and engine speed. The desired spark advance is found from addressing a look-up table in memory according to MAP and RPM.

Once the desired spark advance has been found, the computer generates an output pulse which triggers the electronic ignition circuit. For example, if cylinder 1 or 4 is to be fired, a signal will appear on output a. To fire cylinder 2 or 3, the signal will appear on output b. This signal will cause the ignition circuit to interrupt current flow in half of the primary of the coil. The secondary of the coil will be energized with a polarity which depends upon which half of the primary has been interrupted. For example, if the computer activates output a, coil secondary terminal c will be positive relative to terminal d. Current will flow through diodes D1 and D4 to fire spark plugs 1 and 4. Spark plugs 2 and 3 will not be fired because the secondary voltage causes diodes D2 and D3 to be reverse biased; that is, they block the current flow.

Similarly, whenever the computer generates a pulse on output b, the current through the lower half of the coil primary will be interrupted. This will cause coil secondary terminal d to be positive with respect to terminal c. Current will flow through diodes D2 and D3 to fire spark plugs 2 and 3.

Diodes which have the capability of blocking current flow when reverse biased by high voltages must be used. Such diodes are not particularly cheap, but they are less expensive than the distributor and components which they replace. Also, since all moving mechanical parts are eliminated, this method is likely to be more trouble free.

There are other configurations for distributor-less ignition that have been successfully tested on automobiles and have proven to be technically feasible. Some use two separate ignition coils and others employ electronic switching of coil secondary voltage to the appropriate spark plug.

Transmission Control

The automatic transmission is another important part of the drivetrain that must be controlled. Traditionally, the automatic transmission control system has been hydraulic and pneumatic. However, there are some potential benefits to the electronic control of the automatic transmission.

The engine and transmission work together as a unit to provide the variable torque needed to move the car. If the transmission were under control of the electronic engine control system, then optimum performance for the entire drivetrain could be obtained by coordinating the engine controls and transmission gear ratio. Various experimental programs for electronic control of shifting of the automatic transmission have been tried; however, at this time, none have been used in a production vehicle.

Torque converter lockup
reduces the slip that normally occurs within an automatic transmission by
mechanically locking the
driving pump and the
driven turbine.

One aspect of automatic transmission control, called torque converter lock-up, is likely to be electronic in the near future. The torque converter is a fluid coupling between the engine output shaft and the gear mechanism of the transmission. The fluid coupling is required to decouple the engine and gear system whenever the vehicle is stopped with engine running and a gear selected. In addition, the torque converter provides a useful torque multiplication during acceleration.

The torque converter has one major disadvantage; that is, the pump turns at a slightly higher RPM than the turbine when delivering power to the drive wheels. This phenomenon, called torque converter slip, reduces the efficiency of the drive train since all of the engine output power is not delivered to the drive wheels. For efficiency, it would be better to mechanically lock the driving (pump) and driven (turbine) elements of the torque converter together. Torque converter lock-up is desirable under the combined conditions of relatively low load and relatively low slip as occurs at steady cruise. Under these conditions, it is possible to lock the two elements together by moving a splined gear along the shaft of one element to engage the spline on the other shaft, thereby locking the input and output elements together.

Various schemes have been considered for achieving torque converter lock-up. One such scheme adds this function to the engine control system. *Figure 10-2* is a partial block diagram of an engine control system having torque converter lock-up as one of its functions.

**Figure 10-2.
Torque Converter
Lock-Up Control System
Block Diagram**

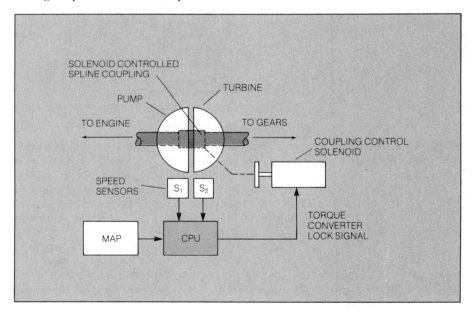

One torque converter lockup system relies on sensors that indicate when conditions are right for converter lockup. Upon a signal from the computer, a solenoid pulls a locking coupling into place between the pump and the turbine.

This system includes sensors S_1 and S_2 which measure the angular speed of the pump and turbine respectively. These sensors can be magnetic-type position sensors such as those discussed in Chapter 5. The computer can calculate slip using the sensor output signals. In addition, the MAP sensor provides an indication of the engine load (i.e. torque requirement). Whenever the manifold pressure and slip satisfy conditions for which lock up is safe, the CPU generates an output to the torque converter lock-up actuator. This can be, for example, a solenoid which slides the locking coupling into place.

It should be noted that torque converter lock up, as for most control functions, does not have to be electronically implemented because it can be accomplished using mechanical, hydraulic and pneumatic controls. However, the electronic lock up has greater flexibility and can be applied in a wider range of vehicle operating conditions.

AUTOMOTIVE SAFETY RELATED ELECTRONIC SYSTEMS

Quite a few interesting concepts pertaining to safety have been considered and brought to different levels of development. Some of these have been initiated by proposed government regulation. Others are only "wishful thinking" and have not been seriously studied. We will present a few of these to discuss technical aspects although the likelihood of production application is very uncertain.

Vehicle Occupant Protection

One such safety concept which has been widely discussed in the news media and which has been strongly pushed by the U.S. government is the "air bag". The air bag is for protection of the occupants of a car in collisions that usually cause severe injury and death, such as the head-on collision. In a head-on collision, the occupants are thrown forward against the dashboard, windshield, and protruding objects, particularly the steering wheel.

The air bag concept takes advantage of the very short time interval from the instant the vehicle hits something until the car occupants start moving forward. For example, for a car traveling at 60 miles per hour striking a fixed object (e.g. tree), a time interval of about 50 to 100 milliseconds (a millisecond is one thousandth of a second) will pass before the occupants start moving forward. During this time, a flexible bag similar to a large balloon is inflated between the occupants and the dash or steering wheel as shown in the sketch of *Figure 10-3*.

Air bags are flexible, inflatable bags that are used in the event of a collision to minimize injuries to the vehicle's occupants. Air bags are inflated under electronic control.

The air bag absorbs the energy of the forward motion of the occupants and prevents contact between them and the rigid portion of the car interior. In fact, tests with dummies which closely simulate the human body and its motion in the car, have shown that the occupants can survive a relatively high speed head-on collision with minimal injury.

**Figure 10-3.
Air Bag Inflation**

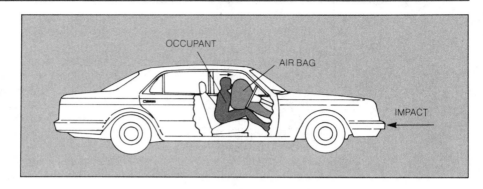

The block diagram for a hypothetical electronic air bag system is shown in *Figure 10-4*. In this scheme, the air bag is folded and placed in a small container in the dashboard. Compressed gas which will be used to inflate the air bag is in another container which is connected to the air bag. The two devices are isolated by a solenoid operated valve which is normally tightly closed to prevent leakage of the compressed gas into the air bag.

The occurrence of a severe impact (e.g. head-on collision) is detected by means of an impact sensor. Normally this impact sensor would be mounted near the front of the vehicle where it would provide earliest possible detection of a head-on collision. In our hypothetical system, this sensor consists of a pendulum and ring electrode switch. The pendulum is held in an open circuit position by a relatively strong spring. Normal motion of the car, including rapid acceleration and hard braking, will not move the pendulum enough to make contact with the ring.

In the event of a collision, an impact sensor sends a signal to a processing circuit opening a valve-operated solenoid. The solenoid allows compressed gas to flow from a tank into the air bag.

**Figure 10-4.
Air Bag Control System**

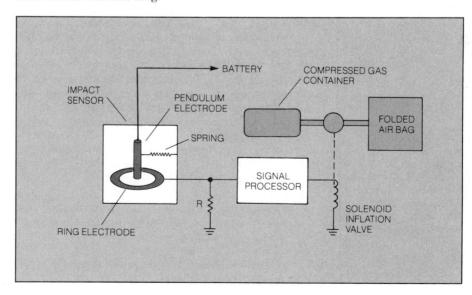

However, the rapid deceleration of an impact causes the pendulum to swing over to contact the ring, thereby closing the switch. The closed switch sends an electric signal to the electronic signal processor. The signal from the impact sensor would be processed (possibly detecting false alarms) and the electrical signal which operates the solenoid valve to inflate the air bag would be generated.

There have been numerous technical problems in developing the air bag system. However, the air bag has been tested and has actually been placed in some production vehicles, although its cost and the lack of customer acceptance caused it to be dropped from production.

A major non-technical issue affecting utilization of the air bag in production vehicles is the liability for air bag malfunction. In spite of highly developed technology, it is impossible to construct any system which makes no errors. If the air bag should inflate by error under normal driving conditions, it could *cause* an accident. If it should fail to inflate as expected in a collision, occupants may be injured. The question of who would be liable for such events has not been resolved; therefore, the future of this system is very uncertain.

Collision Avoidance Radar Warning System

Another interesting safety related electronic system having potential for future automotive application is the anti-collision warning system. An on-board low-power radar system can be used as a sensor for an electronic collision avoidance system to provide warning of a potential collision with an object in the path of the vehicle. As early as 1976, at least one experimental system was reported which could accurately discriminate objects up to distances of about 100 yards. This system gave very few false alarms in actual highway tests.

Collision avoidance radar systems use low-power radar to sense objects and provide warnings of possible collisions.

For an anti-collision warning application, the radar antenna should be mounted on the front of the car and it should project a relatively narrow beam forward. Ideally the antenna for such a system should be in as flat a package as possible and should project a beam which has a width of about 2 to 3 degrees horizontally and about 4 to 5 degrees vertically. Also, if the beam is scanned horizontally for a few degrees, say 2.5 degrees either side of center, false alarms from roadside objects can be reduced. Large objects, such as signs, can reflect the radar beam, particularly on curves, and trigger a false alarm.

In order to test whether a detected object is in the same lane as the radar equipped car traveling around a curve, the radius of the curve must be measured. This can be estimated closely from the front wheel steering angle for an unbanked curve. Given the scanning angle of the radar beam and the curve radius, a computer can quickly perform the calculations to determine whether a reflecting object is in the same lane as the protected car or not.

For the collision warning system, better results can be obtained if the radar transmitter is operated in a pulsed mode rather than a continuous wave mode. In this mode, the transmitter is switched on for a very short time, then it is switched off. During the off time, the receiver "looks for" a reflected signal. If a reflecting object is in the path of the transmitted microwave pulse, then a corresponding pulse will be reflected to the receiver. The round trip time, t, from transmitter to object and back to receiver is proportional to the range, R, to the object as illustrated by *Figure 10-5:*

$$t = 2R/c$$

where

$$c = \text{speed of light (186,000 miles per second)}$$

The radar system has the capability to accurately measure this time to determine the range to the object.

**Figure 10-5.
Range to Object for Anti-
Collision Warning
System**

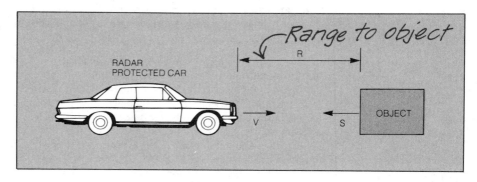

It is possible to measure the vehicle speed, V, by measuring the Doppler frequency shift of the pulsed signal which is reflected by the ground. (The Doppler frequency shift is proportional to the speed of the moving object. The Doppler shift is what causes the pitch of the whistle of a moving train to change as it passes.) This reflection can be discriminated from the object reflection because the ground reflection is at a low angle and a short, fixed range.

The reflection from an object will have a pulse shape which is very nearly identical to the transmitted pulse. As we said, the radar system can detect this object reflection and find R to determine the distance from the vehicle to the object. In addition, the relative speed-of-closure between the car and object can be calculated by adding the vehicle speed, V, from the ground reflected pulses, and the speed of the object, S, which can be determined from the change in range of the object's reflection pulses.

A collision avoidance system compares the time needed for a microwave signal to be reflected from an object to the time needed for a signal to be reflected from the ground. By comparing these times with vehicle speed data, the computer can calculate a "time to impact" value, and sound an alarm if necessary.

A block diagram of an experimental collision warning system is shown in *Figure 10-6*. In this system, the range, R, to the object and the closing speed, V + S, are measured. The computer can perform a number of calculations on this data. For example, the computer can calculate the time-to-collision T:

$$T = R/(V + S)$$

Whenever this time is less than a preset value, a visual and audible warning is generated. The system could also be programmed to release the throttle and apply the brakes if automatic control were desired.

If the two objects are traveling at the same speed in the same direction, then S = −V and T would be infinite. That is, a collision would never occur.

If the object were stationary, then S = 0 and the time to collision would be:

$$T = R/V$$

**Figure 10-6.
Block Diagram of
Collision Avoidance
Warning System**

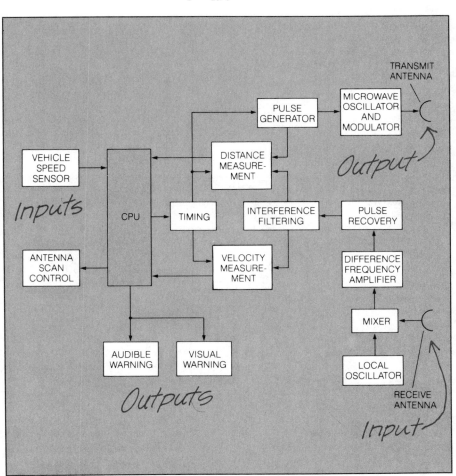

If the object were another moving car approaching the radar equipped car head-on, then the closing speed would be the sum of the two car speeds. In this case, the time to closure would be:

$$T = R/(V + S)$$

This concept has already been considerably refined from its earliest inception. However, there are still some technical problems which must be overcome before this system is ready for production use. Nevertheless, the performance of the experimental systems which already have been tested is impressive. It will be interesting to watch this technology as it improves and to see which, if any, of the present system configurations becomes commercially available.

Low Tire-Pressure Warning System

Another interesting electronic system which may be used on future automobiles is a warning system for low tire pressure that works while the car is in motion. A potentially dangerous situation could be avoided if the driver could be alerted to the fact that a tire has low pressure, particularly if it happens while driving. For example, if a tire were to develop a leak, the driver could be warned in sufficient time to stop the car before control became difficult.

There are several pressure sensor concepts which could be used. A block diagram of a hypothetical system is shown in *Figure 10-7*. In this scheme, a tire pressure sensor, S, continually measures the tire pressure. The signal from the sensor mounted on the rolling tire is coupled by link to the electronic signal processor. Whenever the pressure drops below a critical limit, a warning signal is sent to a display on the instrument panel to indicate which tire has the low pressure.

**Figure 10-7.
Low Tire Pressure
Warning System**

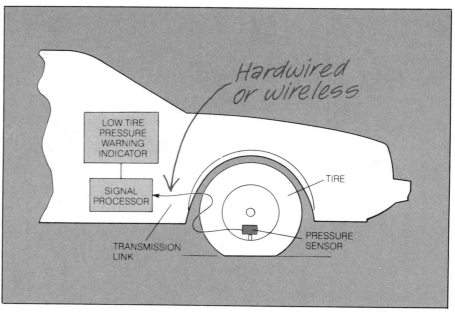

A low tire pressure warning system utilizes a tire mounted pressure sensor. The pressure sensor signals a loss in tire pressure, either by wiring to the axle through slip rings, or radio transmission from a transmitter mounted in the valve stem.

The difficult part of this system is the link from the tire pressure sensor mounted on the rotating tire to the signal processor which is mounted on the body. Several concepts have potential for this link. For example, slip rings which are similar to the brushes on a dc motor could be used. However, this concept would require major modification to the wheel-axle assembly and does not appear to be acceptable at the present time.

Another concept for this link is to use a small radio transmitter mounted on the tire. By using modern solid-state electronic technology, a tiny low-power transmitter could be constructed. The transmitter could be located in a modified tire valve cap and could transmit to a receiver in the wheel well. The distance from the transmitter to the receiver would be only about one foot, so only very low power would be required.

One problem with this method is that electrical power for the transmitter would have to be provided by a self-contained battery. However, the transmitter need only operate for a few seconds and only when the tire pressure falls below a critical level. Therefore, a tiny battery could provide enough power.

This scheme is illustrated schematically for a single tire in *Figure 10-8*. The sensor switch is normally held open by normal tire pressure on a diaphragm mechanically connected to the switch. Low tire pressure allows the spring-loaded switch to close to turn on the microtransmitter. The receiver, which is directly powered by the car battery, receives the transmitted signal and passes it to the signal processor, also directly powered by the car battery. The signal processor then activates a warning lamp for the driver which remains on until the driver resets the warning system by operating a switch on the instrument panel.

**Figure 10-8.
Low Pressure Sensor
Concept**

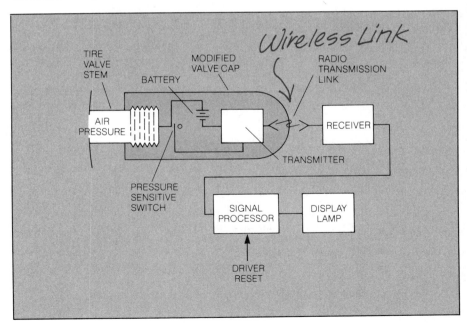

One reason for using a signal processing unit is the relatively short life of the transmitter battery. The transmitter will remain on until the low pressure condition is corrected or until the battery runs down. By using a signal processor, the low pressure status can be stored in memory so the warning will still be given even if the transmitter quits operating. The need for this feature could arise if the pressure dropped while the car was parked. By storing the status, the system would warn the driver as soon as the ignition was turned on.

Many concepts have been proposed for a low tire-pressure warning system. The future of such a system will be limited largely by its cost.

AUTOMOTIVE INSTRUMENTATION

The reduced cost of VLSI and microprocessor electronics are resulting in advanced instrumentation and the use of voice synthesis in warning systems.

It is very likely that some interesting advances in automotive instrumentation are forthcoming. Included in these advances are: functions that are not presently available; new display forms including audible messages by synthesized speech; and interactive communication between the driver and the instrumentation. These advances will come about partly because of increased capability at reduced cost for modern solid-state circuits, particularly microprocessors and microcomputers.

One of the important functions which an all electronic instrumentation system can have in future automobiles is continuous diagnosis of other on-board electronic systems. In particular, the future computer-based electronics instrumentation may perform diagnostic tests on the electronic engine control system. This instrumentation system might display major system faults and even recommend repair actions.

Another function which might be improved in the instrumentation system is the trip computer function. The system probably will be highly interactive; that is, the driver will communicate with the computer through a keyboard, or maybe even by voice.

The full capabilities of such a system are limited more by human imagination and cost than technology. Most of the technology for the systems discussed is available now and can be packaged small enough for automotive use. However, in a highly competitive industry where the use of every screw is analyzed for cost-effectiveness, the cost of these systems still limits their use in production vehicles.

CRT Display

One of the exciting possibilities for future automotive electronic systems is in the display device. The possibility of displaying alphanumeric data to the driver via a cathode ray tube (CRT) has been considered. In fact, some CRT display systems are already being used in automobiles, particularly police cars, for mobile data terminals. The CRT has been used for years to display data in airport terminals and computer terminals. Using the CRT for an automotive display is easy when the automotive instrumentation is computerized. The major problem is the relatively large space requirement in the already crowded instrument panel. Also, the CRT and its required support electronics are relatively expensive. Even these problems soon may be overcome as development of the flat CRT and similar display devices continues.

No other visual display device presently available has the capability and information rate of the CRT. However, this tremendous information display capability could create a problem if data is displayed at high rates to the driver. In its normal installation, the CRT could distract the driver from looking at the road. However, there is a method to display the data without having the driver look away from the road. This method, developed for use in airplanes, is known as the head-up display (HUD).

Figure 10-9 shows the concept of a HUD. In this scheme, the information which is to be displayed appears on a CRT which is mounted as shown. A partially reflecting mirror is positioned above the instrument panel in the driver's line-of-sight of the road. The driver looks through this mirror at the road.

In normal driving, the driver looks through this mirror at the road. Information to be displayed appears on the face of the CRT upside-down and the image is reflected by the partially reflecting mirror to the driver right-side up. He can read this data from the HUD without moving his head from the position for viewing the road.

The CRT when combined with a partially-reflective mirror, results in a HUD. Information is displayed on the CRT in the form of a reversed image. The image is reflected by the mirror, and viewed normally by the driver.

**Figure 10-9.
HUD Display**

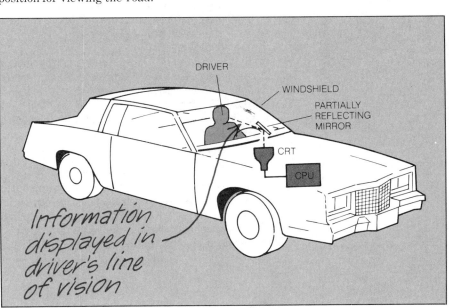

The brightness of this display would have to be adjusted so that it is compatible with ambient light. The brightness of this data image should never be so great that it inhibits the driver's view of the road, but it must be bright enough to be visible in all ambient lighting conditions. Fortunately, the CRT brightness can be automatically controlled by electronic circuits to accommodate a wide range of light levels.

Speech Synthesis

One really exciting new display device, audible display by synthesized speech, has great potential for future automotive electronic instrumentation. Important safety or trip-related messages could be given audibly so the driver doesn't have to look away from the road because the display is via electronically generated artificial speech. In addition to its normal function of generating visual display outputs, the computer generates an electrical wave form which is approximately the same as a human voice speaking the appropriate message. The voice quality of some types of speech synthesis is often quite natural and closely similar to human speech.

Speech synthesizers use phoneme synthesis, a method of imitating basic sounds used to build speech. Computers rely on an inventory of phonemes to build the words for various automotive warning messages.

There are several major categories of speech synthesis which have been studied experimentally. Of these, the phoneme synthesis is probably the most sophisticated. A phoneme is a basic sound which is used to build speech. By having an inventory of these sounds in computer memory and by having the capability to generate each phoneme sound, virtually any word can be constructed by the computer in a manner similar to the way the human voice does. Of course, the electrical signal produced by the computer is converted to sound by a loudspeaker.

Synthesized speech is already being used in learning aids such as the Texas Instruments *Speak and Spell*™ and *Home Computer*. It is also being used in appliances such as a microwave oven. It seems very likely that the voice synthesis display will shortly become available in some production cars. Some auto experts believe that this display method will become widespread before 1990.

MULTIPLEXING IN AUTOMOBILES

One of the high-cost items in building and servicing vehicles is the electrical wiring. Wires of varying length and diameter form the interconnection link between each electrical/electronic component in the vehicle. Virtually the entire electrical wiring for a car is made up in the form of a complex, expensive cable assembly called a "harness". Building and installing the harness requires manual assembly and is time consuming. The increased use of electrical and electronic devices has significantly increased the number of wires in the harness.

Sensor Multiplexing

The use of microprocessors for computer engine control, instrumentation computer, etc., offers the possibility of significantly reducing the complexity of the harness. For example, consider the engine control system. In the present configuration, each sensor and actuator has a separate wire connection to the CPU. However, each sensor only communicates periodically with the computer for a short time interval during sampling.

It is possible to connect all the sensors to the CPU with only a single wire (with ground return, of course). This wire, which can be called a data bus, provides the communication link between all the sensors and the CPU. Each sensor would have exclusive use of this bus to send data (i.e. measurement of the associated engine variable/parameter) during its time slot. A separate time slot would be provided for each sensor.

This process of selectively assigning the data bus exclusively to a specific sensor during its time slot is called time division multiplexing (or sometimes just multiplexing—MUX). Recall that we discussed multiplexing as a data selector for the CPU input and output.

We can understand the operation of time division multiplexing of the data bus by referring to the system block diagram in *Figure 10-10*. The CPU controls the use of the data bus by signaling each sensor through a transmitter/ receiver (T/R) unit. Whenever the CPU requires data from any sensor, it sends a coded message on the bus which is connected to all T/R units.

Sensor multiplexing can reduce the necessary wiring in an electrical harness, by using time division multiplexing.

**Figure 10-10.
Sensor Multiplexing
Block Diagram**

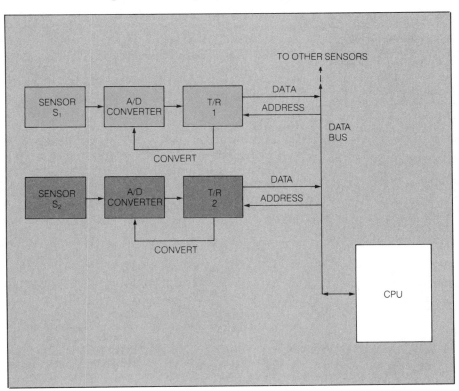

However, the message consists of a sequence of binary voltage pulses which are coded for the particular T/R unit. A T/R unit responds only to one particular sequence of pulses which can be thought of as the address for that unit.

Whenever a T/R unit recognizes its address, it activates an analog to digital converter. The sensor's analog output at this instant is converted to a digital binary number as we've discussed before. This number and the T/R unit's address are placed on the data bus and the CPU reads the data. The T/R unit's address is included so the CPU can identify the source of the data. Thus, the CPU interrogates a particular sensor and then receives the measurement data from the sensor on the data bus. The CPU then sends out the address of the next T/R unit whose sensor is to be sampled.

Control Signal Multiplexing

It also is possible to multiplex control signals to control switching of electrical power. Electrical power must be switched to lights, electric motors, solenoids and other devices.

The system for multiplexing electrical power control signals around the vehicle requires two busses: one carrying battery power and one carrying control signals. *Figure 10-11* is a block diagram of such a multiplexing system. In a system of this type, a remote switch is required at each component, C, which is to be operated. The remote switch applies battery power to the component when activated by a receiver module, RM. The receiver module is activated by a command from the CPU which is transmitted along the control signal bus.

This control signal bus operates very much like the sensor data bus described in the multiplexed engine control system. The particular component which is to be switched is initially selected by switches operated by the driver. (Of course, these switches can be multiplexed at the input of the CPU.) The CPU sends an RM address as a sequence of binary pulses along the control signal bus. Each receiver module responds only to one particular address. Whenever the CPU is to turn on or off a given component, it transmits the coded address and command to the corresponding RM. When the RM receives its particular code, it operates the corresponding switch, either applying battery power or removing battery power depending upon the command which is transmitted by the CPU.

Each sensor in a multiplexed system sends its individual data over a common bus. The computer identifies the sensor by signalling each sensor with a unique address.

A multiplexed system can also control switching of electrical power for lights, motors, and similar devices. Each RM would switch power to the appropriate device in response to a CPU command.

**Figure 10-11.
Control Signal
Multiplexing Block
Diagram**

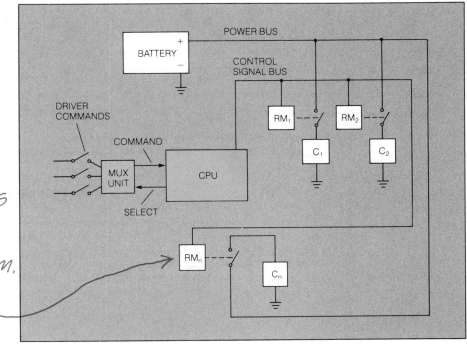

Receiver module activates its related component or subsystem.

Fiber Optics

Signal busses using fiber optics would transmit data and control signals in the form of light pulses along thin fiber cables. Such systems would be relatively immune from noise interference.

It is possible, maybe even desirable, to use an optical fiber[1] for the signal bus. For such a system, the address voltage pulses from the CPU are converted to corresponding pulses of light which are transmitted over an optical fiber. An optical fiber, which is also known as a light pipe, consists of a thin transparent cylinder of light conducting glass about the size of a human hair. Light will follow the "light pipe" along its entire path, even around corners just as electricity follows the path of wire. A big advantage of the optical fiber signal bus for automotive use is that external electrical noise doesn't interfere with the transmitted signal. The high voltage pulses in the ignition circuit, which are a major source of interference in automotive electronic systems, will not affect the signals traveling on the optical signal bus.

For such a system, each component has an RM which has an optical detector coupled to the signal bus. Each detector receives the light pulses which are sent along the bus. Whenever the correct sequence (i.e. address) is received at the RM, the corresponding switch is either closed or opened.

A variety of multiplexing systems have been experimentally studied. It seems very likely that one form or another of multiplex system will be used in the near future.

[1] L.B. and B.R. Masten, *Understanding Optronics*, Texas Instruments Incorporated, Dallas, Texas, 1981

SUMMARY

In this chapter, we have tried to describe in general terms some of the future applications of electronics in the automobile. This survey has not tried to forecast when (or even if) any particular system will appear in production cars. Rather it has tried to explain each system concept and to identify some of the strengths and weaknesses of each.

Broadly speaking, it can reasonably be assumed that electronic systems will be used more and more in cars. To some extent, electronics will replace existing functions which are implemented mechanically or pneumatically/hydraulically. However, it is more likely that new functions will be added to provide better control, more safety features, and more driver convenience.

Electronic systems will continue to be attractive to automobile designers because of the low cost per function, because of the great flexibility for system change at low cost, and because of the greater functional capability of electronic systems for control and instrumentation. It is further likely that many future automotive electronic systems will be microprocessor based.

With the understanding of automotive electronics provided by this book, the reader should be able to analyze and understand the new systems as they become available.

Quiz for Chapter 10

1. In a distributor-less ignition,
   a. the distributor is replaced by a rotor
   b. 2 spark plugs of a 4 cylinder engine may be simultaneously fired
   c. the rotor switches high voltage
   d. there is no coil

2. Torque converter lock-up refers to:
   a. a key operated safety lock for the transmission
   b. an all-electronic torque converter
   c. a mechanism for mechanically linking the pump and turbine of a torque converter
   d. none of the above

3. An air bag is
   a. a mechanism for occupant protection in a car
   b. a container for use in case of airsickness
   c. an impact sensor
   d. all of the above

4. One concept for automotive collision avoidance involves
   a. braking rapidly in dangerous situations
   b. measuring the round-trip time of a radar pulse from protected car to collision object
   c. aircraft surveillance of highways
   d. wheel speed sensors

5. Doppler shift has potential automotive application for
   a. measuring the speed of passing trains
   b. automatic gear changing
   c. measuring vehicle speed over the road
   d. none of the above

6. The major problem associated with a practical low tire pressure sensor is
   a. developing a means of getting the electrical signal from the rotating sensor to the car body
   b. developing a sensor which can measure relative pressure
   c. absolute pressure calibration
   d. measuring tire pressure when the vehicle is stopped

7. A CRT has potential automotive application for:
   a. controlling vehicle motion
   b. recording vehicle transient motion
   c. monitoring entertainment systems
   d. displaying information to the driver

8. The term HUD refers to
   a. housing and urban development
   b. head-up display
   c. heads-up driver
   d. none of the above

9. Speech synthesis is
   a. a system which automatically recognizes human speech
   b. an automatic check-book balancing system
   c. a visual display of speech waveforms
   d. a means of electronically generating human speech

10. An optical fiber is
   a. a tiny beam of light
   b. an optical waveguide which is often called a light pipe
   c. an optical switch
   d. none of the above

Glossary

A/F: See Air/fuel Ratio.

Accumulator: The basic work register of a computer.

Actuator: A device which performs an action in response to an electrical signal.

A/D (also ADC): Analog to digital converter; a device which produces a number proportional to the analog voltage level input.

Air/Fuel Ratio: The ratio of the mass of air to the mass of fuel drawn into a cylinder.

Analog Circuits: Electronic circuits which amplify, reduce or otherwise alter a voltage signal which is a smooth or continuous copy of some physical quantity.

Assembly Language: An abbreviated computer language which humans can use to program computers. Assembly language eventually is converted to machine language so that a computer can understand it.

BDC: Bottom dead center; the extreme lowest position of the piston during its stroke.

Bit: A binary digit; the smallest piece of data a computer can manipulate.

Block Diagram: A system diagram which shows all of the major parts and their interconnections.

BSCO: Brake specific CO; the ratio of the rate at which carbon monoxide leaves the exhaust pipe to the brake horsepower.

BSFC: Brake specific fuel consumption; the ratio of the rate at which fuel is flowing into an engine to the brake horsepower being generated.

BSHC: Brake specific HC; the ratio of the rate at which hydrocarbons leave the exhaust pipe to the brake horsepower.

$BSNO_x$: Brake specific NO_x; the ratio of the rate at which oxides of nitrogen leave the exhaust pipe to the brake horsepower.

Byte: 8 bits dealt with together.

CAFE: Corporate-Average-Fuel-Economy; The government mandated fuel economy which is averaged over the production for a year for any given manufacturer.

Capacitor: An electronic device which stores charge.

Catalyst: A material which speeds up or stimulates a chemical reaction.

Catalytic Converter: A device which enhances certain chemical reactions which help to reduce the levels of undesirable exhaust gases.

Closed Loop Fuel Control: A mode where input air/fuel ratio is controlled by metering fuel in response to the rich-lean indications from an exhaust gas oxygen sensor.

CO: Carbon monoxide; an undesirable chemical combustion product due to imperfect combustion.

Combinational Logic: Logic circuits whose outputs depend only on the present logic inputs.

Combustion: The burning of the fuel-air mixture in the cylinder.

Comparator, Analog: An electronic device which compares the voltages applied to its inputs.

Compression Ratio: The ratio of the cylinder volume at BDC to the volume at TDC.

Control Variable: The plant inputs and outputs which a control system manipulates and measures to properly control it.

Conversion Efficiency (Catalytic Converter): The efficiency with which undesirable exhaust gases are reduced to acceptable levels or are converted to desirable gases.

CPU: Central processing unit; the calculator portion of a computer.

Cutoff: A transistor operating mode where very little current flows between the collector and emitter.

D/A (also DAC): Digital to analog converter; a device which produces a voltage which is proportional to the digit input number.

Damping Coefficient: A parameter which affects a system's time repsonse by making it more or less sluggish.

DEMUX: Demultiplexer; a type of electronic switch used to select one of several output lines.

Diesel: A class of internal combustion engine in which combustion is initiated by the high temperature of the compressed air in the cylinder rather than an electrical spark.

Digital Circuits: Electronic circuits whose outputs can charge only at specific instances and between a limited number of different voltages.

Diode: A semiconductor device which acts like a current check valve.

Display: A device which indicates in human readable form the result of measurement of some variable.

Drivetrain: The combination of mechanisms connecting the engine to the driving wheels including transmission, driveshaft, and differential.

Dwell: The time that current flows through the primary circuit of the ignition coil for each spark generation.

Dynamometer: A device for loading the engine and measuring engine performance.

EGO: Exhaust gas oxygen; the concentration of oxygen in the exhaust of an engine. An EGO sensor is used in closed loop fuel control systems to indicate rich or lean A/F.

EGR: Exhaust gas recirculation; a procedure in which a portion of exhaust is introduced into the intake of an engine.

Electronic Carburetor: A fuel metering actuator in which the air/fuel ratio is controlled by continual variations of the metering rod position in response to an electronic control signal.

Engine Calibration: The values for air/fuel, spark advance and EGR at any operating condition.

Engine Crankshaft Position: The angular position of the crankshaft relative to a reference point.

Engine Mapping: A procedure of experimentally determining the performance of an engine at selected operating points and recording the results.

Equivalence Ratio: Actual air/fuel ratio divided by the air/fuel ratio at stoichiometry.

Evaporative Emissions: Evaporated fuel from the carburetor or fuel system which mixes with the surrounding air.

Foot-Pound: A unit of torque corresponding to a force of one pound acting on a one foot lever arm.

Frequency Response: A graph of a system's response to different frequency input signals.

Gain: The ratio of a system's output magnitude to its input magnitude.

HC: Hydrocarbon chemicals, such as gasoline, formed by the union of carbon and hydrogen.

Ignition Timing: The time of occurrence of ignition measured in degrees of crankshaft rotation relative to TDC.

Inductor: A magnetic device which stores energy in a magnetic field produced by current flowing in it.

Instrumentation: Apparatus (often electronic) which is used for measurement or control, and for display of measurements or conditions.

Integral Amplifier: A control system component whose voltage output changes at a rate proportional to its input voltage.

Integrated Circuit: A semiconductor device which contains many circuit functions on a single chip.

Interrupts: An efficient method of quickly requesting a computer's attention to a particular external event.

Lead Term: A control system component which anticipates future inputs based on the current signal trend.

Limit Cycle: A mode of control system operation in which the controlled variable cycles between extreme limits with the average near the desired value.

Linear Region: A transistor operating mode where the collector current is proportional to the base current.

Logic Circuits: Digital electronic circuits which perform logical operations such as NOT, AND, OR and combinations of these.

Look-Up Table: A table in computer memory which is used to convert an important value into a related value from the table.

MAP: Manifold absolute pressure; the absolute pressure in the intake manifold of an engine.

Mathematical Model: A mathematical equation which can be used to numerically compute a system's response to a particular input.

Microcomputer: A small computer which uses an integrated circuit which contains a central processing unit and other control electronics.

MUX: Multiplexer; a type of electronic switch used to select one of several input lines.

NO$_x$: The various oxides of nitrogen.

Op-Code: A number which a computer recognizes as an instruction.

Open Loop Fuel Control: A mode where engine input air/fuel ratio is controlled by measuring the mass of input air and adding the proper mass of fuel to obtain a 14.7 to 1 ratio.

Operational Amplifier: A standard analog building block with two inputs, one output and a very high voltage gain.

Optimal Damping: The damping which produces the very best time response.

Peripheral: An external input-output device which is connected to a computer.

Phase Shift: A measure of the delay in degrees between the time a signal enters a system and the time it shows up at the output as a fraction of a full cycle of 360°.

Plant: A system which is to be controlled.

Proportional Amplifier: A control system component which produces a control output proportional to its input.

Qualitative Analysis: A study which reveals how a system works.

Quantitative Analysis: A study which determines how well a system performs.

RAM: Random access memory; read/write memory.

Random Error: A measurement error which is neither predictable nor correctable, but has some statistical nature to it.

ROM: Read only memory; permanent memory used to store permanent programs.

RPM: Revolutions per minute; the angular speed of rotation of the crankshaft of an engine or other rotating shaft.

Sample and Hold: The act of measuring a voltage at a particular time and storing that voltage until a new sample is taken.

Sampling: The act of periodically collecting or providing information about a particular process.

Semiconductor: A material which is neither a good conductor nor a good insulator.

Sensor: An energy conversion device which measures some physical quantity and converts it to an electrical quantity.

Sequential Logic: Logic circuits whose output depends on the particular sequence of the input logic signals.

SI Engine: Abbreviation for spark ignited, gasoline fueled, piston type, internal combustion engine.

Signal Processing: The alteration of an electrical signal by electronic circuitry; used to reduce the effects of systematic and random errors.

Skid: A condition in which the tires are sliding over the road surface rather than rolling; usually associated with braking.

Slip: The ratio of the angular speed of the driving element to the angular speed of the driven element of a torque converter; also, the condition in which a driven tire loses traction so that the driving torque does not produce vehicle motion.

Software: The computer program instructions used to tell a computer what to do.

Spark Advance: The number of degrees of crankshaft rotation before TDC where the spark plug is fired. (See ignition timing.)

Spark Timing: The process of firing the spark plugs at the proper moment to ignite the combustible mixture in the engine cylinders.

Stoichiometry: The air/fuel ratio for perfect combustion; it enables exactly all of the fuel to burn using exactly all of the oxygen in the air.

System: A collection of interacting parts.

Systematic Error: A measurement error in an instrumentation system which is predictable and correctable.

TBFI: Throttle-body-fuel-injector; a fuel metering actuator in which the air/fuel ratio is controlled by injecting precisely controlled spurts of fuel into the air stream entering the intake manifold.

TDC: Top dead center; the extreme highest point of the piston during its stroke.

Throttle Angle: The angle between the throttle plate and a reference line; engine speed increases as the angle increases.

Torque Converter: A form of fluid coupling used in an automatic transmission which acts like a torque amplifier.

Torque: The twisting force of the crankshaft or other driving shaft.

Transfer Function: A mathematical equation which, when graphed, produces a system's frequency response plot.

Transistor: An active semiconductor device which operates like a current valve.

Transport Delay: The time required for a given mass of fuel and air to travel from the intake manifold through the engine to the EGO sensor in the exhaust manifold.

Volumetric Efficiency: The pumping efficiency of the engine as air is drawn into the cylinders.

Index

Answers to Quizzes

Chapter 1
1. c
2. a
3. b
4. d
5. b
6. b
7. c
8. b
9. a
10. a
11. a
12. d

Chapter 2
1. e
2. c
3. b
4. a
5. b
6. c
7. b
8. d
9. d
10. d

Chapter 3
1. a
2. c
3. b
4. a
5. c
6. b
7. d
8. e
9. d
10. e
11. d
12. d
13. b
14. a
15. d
16. c
17. b
18. d
19. c
20. a

Chapter 4
1. a
2. d
3. c
4. a
5. e
6. c
7. d
8. c
9. b
10. a
11. b
12. c
13. d
14. b
15. a
16. b
17. b
18. d
19. d
20. d

Chapter 5
1. b
2. c
3. a
4. b
5. b
6. d
7. c
8. a
9. b
10. a
11. b
12. d
13. a
14. a
15. b
16. d
17. a
18. c
19. b
20. d

Chapter 6
1. c
2. c
3. d
4. b
5. b
6. a
7. c
8. c
9. a
10. c
11. c
12. a
13. c
14. b
15. a
16. d
17. a
18. b
19. c
20. a

Chapter 7
1. c
2. d
3. a
4. d
5. b
6. b
7. c
8. c
9. b
10. d

Chapter 8
1. b
2. c
3. a
4. b
5. a
6. b
7. c
8. b
9. c
10. c

Chapter 9
1. a
2. c
3. c
4. c
5. a
6. c
7. c
8. a
9. c
10. d
11. a
12. a
13. a
14. b
15. c
16. c
17. b
18. a
19. c
20. c

Chapter 10
1. b
2. c
3. a
4. b
5. c
6. a
7. d
8. b
9. d
10. b